Forward-Looking Decision Making

The Gorman Lectures in Economics

Series Editor, Richard Blundell

A series statement appears at the back of the book

Forward-Looking Decision Making

Dynamic-Programming Models Applied to Health, Risk, Employment, and Financial Stability

Robert E. Hall

Princeton University Press

Princeton and Oxford

Copyright © 2010 by Princeton University Press

Published by Princeton University Press,
41 William Street, Princeton, New Jersey 08540

In the United Kingdom: Princeton University Press,
6 Oxford Street, Woodstock, Oxfordshire OX20 1TW

Library of Congress Cataloging-in-Publication Data

Hall, Robert Ernest, 1943–
    Forward-looking decision making : dynamic
    programming models applied to health, risk, employment,
    and financial stability / Robert E. Hall.
        p. cm. – (The Gorman lectures in economics)
    Includes bibliographical references and index.
    ISBN 978-0-691-14242-5 (alk. paper)
    1. Households–Decision making–Econometric models.
    2. Families–Decision making–Econometric models.
    3. Decision making–Econometric models. I. Title.

HB820.H35 2010
330.01′5195–dc22   2009042465

British Library Cataloging-in-Publication Data is available

This book has been composed in LucidaBright using TEX
Typeset and copyedited by T&T Productions Ltd, London

Printed on acid-free paper. ∞
press.princeton.edu

Printed in the United States of America

10 9 8 7 6 5 4 3 2 1

Contents

Foreword

Forward-looking behavior is at the heart of economics. Choices over savings, occupations, earnings, investments, etc., all require forward planning under uncertainty. Just how individuals and firms go about doing this and how, as economists, we should best model what they do are key to our understanding of economic behavior and are the core concern of this book. Few people have had such an impact on the development of these aspects of economics as has Robert Hall.

This volume makes a compelling case for the use of the dynamic-programming approach to modeling choices across a wide range of economic decision making. Not just forward-looking consumption and labor market decisions, but also health-care choices, where the payoff is measured in terms of future quality of life. Entrepreneurial startup decision making under uncertainty is carefully examined too and a coherent framework for examining moral hazard issues in the provision of insurance to the risks faced on uncertain investments is provided, using this to show the value of venture investors.

The reader is introduced to the dynamic-programming approach to modeling forward-looking behavior in an accessible but rigorous way, emphasizing the relationship to the choices being made by the decision maker and the key issues the researcher will face in practical implementation. The key preference parameters required for modeling choices are described as well as how to recover them from empirical analysis in a fashion that is consistent with the forward decision making framework.

We were extremely fortunate to have Robert Hall present these ideas in the Gorman Lecture series and it is wonderful to have them now brought together in one volume.

Richard Blundell
University College London
Institute for Fiscal Studies

Preface

When Richard Blundell asked me to give the Gorman Lectures, I instantly accepted, but it raised the question of whether there was any common theme in my current research. In addition to my career-long struggle to understand fluctuations in the labor market, my recent work has looked at aggregate health spending, entrepreneurship, and financial instability. It dawned on me that I had approached this heterogeneous collection of subjects with a common analytical core—a family or personal dynamic program. In all of the models, people balance the present against the future. So the unifying feature of the chapters in this volume is a modeling approach.

The first chapter quickly reviews ideas about dynamic programs, a topic likely to be familiar to many readers. I make some suggestions about numerical solution and the representation of solved models as Markov processes that may be novel to practitioners. Chapter 2 surveys recent research of the parameters of preferences that often appear in personal dynamic programs: the intertemporal elasticity of substitution, the Frisch elasticity of labor supply, and the Frisch cross-elasticity, a topic that has gained a lot of attention recently.

Chapter 3 covers the first substantive model, one of aggregate health spending, developed jointly with Charles Jones. Here families divide current resources between consumption and health care; the payoff to health care is lowered mortality and increased quality of life. Formulating the dynamic program requires close attention to modeling preferences for length of life. We also use econometric estimates of the relation between health spending and life extension. We argue that the rise in the share of GDP devoted to health is consistent with optimal choice and is not an automatic sign of waste. The dynamic program has a large number of state variables, but is analytically benign because the value function is linear in the state variables.

Throughout my work using family dynamic programs, I have been aware that I was part of a large group of applied economists using Richard Bellman's useful tool. We teach dynamic programming to all first-year Ph.D. students, and, not surprisingly, they reach for it to solve many problems of intertemporal choice. Chapter 4 discusses a fine example of work on an important policy issue by two leading applied economists, Jeffrey Brown and Amy Finkelstein. They study the adverse effects of the availability of public insurance for nursing home and other long-term care on the demand for private insurance for that risk. Their dynamic program places a value on the gains to families if private insurance could be used in conjunction with public benefits, a practice currently barred in the United States.

Chapter 5 discusses my current thinking about fluctuations in the labor market. The model seeks to explain a finding that has troubled me and many other macroeconomists: the apparent inefficiency that occurs in a recession, when idleness among workers seems to signify a low marginal value of time, but measures of the marginal product of labor do not fall that low. A family dynamic program, using preferences from chapter 2, seems to track consumption and weekly hours of work quite reasonably, so the remaining issue is movements in unemployment. Here I incorporate ideas from the recent literature on search-and-matching frictions.

Chapter 6 concerns family economics in two senses, because it is joint work with my economist wife, Susan Woodward, who has assembled a rich body of data on outcomes for entrepreneurs who are in venture-backed startup companies. The dynamic program gives a unified treatment to two kinds of risk facing the entrepreneurs—the amount of the eventual financial reward and the effort needed to achieve the reward. Insurance against the risk would be hugely beneficial to entrepreneurs but is impractical because of moral hazard and adverse selection. Venture investors limit the payoff to a share of the ultimate value of the company to give a powerful incentive to the entrepreneur and to avoid giving money away to people who do not have commercially valuable ideas.

Chapter 7 responds to current financial turmoil by looking at one issue: the potential contribution to instability from government guarantees of private debt. The effect of a guarantee is to create a joint incentive for borrowers and lenders to harvest the guarantee, which enriches the borrower directly and thus influences the terms of the deal between the borrower and lender. When financial institutions are facing substantial probabilities of default, families consume more, because the government may bail out their extra consumption. The rise in consumption drives down interest rates. Some of the features of financial crises may result from the factors considered in the model.

I am pleased to contribute to the activities that remember and honor Terence Gorman. He was a major influence on my development as an economist, though I spent little time with him personally. Unlike many of his colleagues, he was not inclined to spend much time touring the colonies. Reading the remarks at Terence's memorial in 2003, including those of my thesis adviser, Robert Solow, reminded me of how thoroughly I was under Terence's influence. His favorite topics of capital aggregation and duality were much in the air when I was a graduate student at MIT in the mid 1960s. My first sole-authored professional publication—in the *Review of Economic Studies*, naturally—dealt with exactly these topics.

I was lucky enough to spend the amazing years 1967 to 1970 at Berkeley, and take part in a revolution in thinking about household and production economics. Dan McFadden taught me modern theory in this area, much under Terence's influence. Frank Hahn's remark at the memorial about Terence's remaking of consumer theory aptly captures what I learned in Berkeley: "It banished nasty bordered Hessians... it substituted the much deeper and simpler notions of concavity and convexity, and made this branch of theory really quite beautiful."

I am grateful to University College London for hosting me for a week in connection with the lectures and providing many opportunities to discuss the work with many of the world's experts in the areas of my interests. I thank, as ever, the Hoover Institution for supporting my research over the past thirty years and Charlotte Pace for running my office and making many improvements to the manuscript. The National Bureau of Economic Research provided numerous opportunities to air the work and receive needed criticism.

My website (Google "Bob Hall") contains a large amount of supporting material for the research discussed in this book, including data and Matlab programs.

Chapters 2 and 5 draw from my paper, "Reconciling cyclical movements in the marginal value of time and the marginal product of labor," *Journal of Political Economy*, April 2009, pp. 281–323, used with permission of the publisher, © 2009 by the University of Chicago. Chapter 3 draws from my paper with Charles I. Jones, "The value of life and the rise in health spending," *Quarterly Journal of Economics*, February 2007, pp. 39–72, used with permission of the publisher, © 2007 by the President and Fellows of Harvard College and the Massachusetts Institute of Technology. Figure 4.1 is used with permission from Jeffrey R. Brown and Amy Finkelstein from their paper, "The interaction of public and private insurance: Medicaid and the long-term care insurance market," *American Economic Review*, September 2008, pp. 1083–102, © 2008 by the American Economic Association. Chapter 6 draws from my paper with Susan Woodward, "The burden of the nondiversifiable risk of entrepreneurship," *American Economic Review*, forthcoming, used with permission of the publisher, © 2009 by the American Economic Association.

Forward-Looking Decision Making

1

Basic Analysis of Forward-Looking Decision Making

Individuals and families make the key decisions that determine the future of the economy. The decisions involve balancing current sacrifice against future benefits. People decide how much to invest in health care, exercise, and good diet, and so determine their longevity and future satisfaction. They make choices about jobs that determine employment and unemployment levels. Their investment decisions are at the heart of some issues of financial stability.

1.1 The Dynamic Program

Economists have gravitated to the dynamic program as the workhorse model of the way that people balance the present against the future. The dynamic program is one of the two tools economists tend to reach for when solving problems of optimization over time. The other is the Euler equation. The two approaches are perfectly equivalent: if a problem is susceptible to solution by a dynamic program, it is susceptible to an Euler equation solution, and vice versa. The class of models suited to either method have the property that the trade-off between this year and next year—the marginal rate of substitution—depends only on consumption this year and next year. Utility functions with this property are *additively separable*. They have the form,

$$\sum_t \beta^t u(x_t), \tag{1.1}$$

where x_t is a vector of things the family cares about.

In words, the idea of a dynamic program is to summarize the entire future in a *value function*, which shows how much lifetime utility the family will enjoy starting next year based on the resources for future spending left after this year. People makes choices about this year by balancing the immediate marginal benefits from using resources this year against the marginal value of the remaining resources as of next year, according to the value function. A dynamic program uses backward recursion. Start with a known or assumed value function for some distant year. Find the value function in the previous year by solving the year-to-year balancing problem for all possible levels of resources in that year. Keep moving backward in time until you reach a year from now. Finally, solve this year's optimization by taking actual resources and solving the balancing problem for this year's utility and next year's value function. The dynamic-programming approach is conceptually simple and numerically robust. A famous book by Stokey and Lucas (1989) helped persuade economists of the virtues of dynamic programming for recursive problems.

The Euler equation approach considers a possible choice this year and then uses the marginal rate of substitution to map it into a choice next year. Apply

this as a forward recursion, to see where the immediate choice leads to at some distant future date. Again,
as in the dynamic program, you need to have some
concept of a distant target—for example, the family
cannot be deeply in debt at that time. Keep trying out
initial choices to find the one where the family meets
its distant target. Though this approach has some ap-
peal in explaining dynamic models, it fails completely
as a way to solve models numerically. The Euler equation is numerically unstable. Good methods exist for
dealing with the instability, but are rarely used in
modern dynamic economics because the dynamic program approach works so well. The instability of the
Euler equation creates conceptual confusion as well,
because one might fall into the trap of thinking that
the aberrant paths that do not reach the target might
describe actual behavior.

All the models in this book rest on dynamic programming. At the beginning of the year, the family
inherits some *state variables* as a result of choices
made last year and events during that year. These
could be wealth, health status, employment status, or
debt. The family picks values of *choice variables*: consumption, health spending, job search, or borrowing.
A *law of motion* shows how the inherited values of
the state variables, the values of the choice variables,
and events during the year map into next year's values of the state variables. The events include financial
returns, health outcomes, job loss, and fluctuations in
earnings.

The *Bellman equation* describes the condition for
recursive optimization:

$$V_t(s_t) = \max_{x_t}(u_t(s_t, x_t) + \beta \, \mathbb{E}_t \, V_{t+1}(s_{t+1})). \qquad (1.2)$$

Here $V(s)$ is the value function, which assigns lifetime expected utility based on the state variables, s. The choice variables are in the vector x_t. The function $u(s, x)$ is the *period utility* that describes the flow of satisfaction that the family receives when choosing x given s. The general law of motion is

$$s_{t+1} = f_t(s_t, x_t, \epsilon_t). \tag{1.3}$$

Here ϵ_t describes the random events that occur during year t. Anything about year t that is known in advance can be built into the function f_t.

As a simple example, consider the standard life-cycle saving model. The single state variable is wealth, W_t. The only choice variable is current consumption, c_t. The Bellman equation is

$$U_t(W_t) = \max_{c_t}(u(c_t) + \beta \, \mathbb{E}_{\epsilon_t} U_{t+1}(W_{t+1})) \tag{1.4}$$

and the law of motion for wealth is

$$W_{t+1} = (1 + r_t)(W_t - c_t + y_t + \epsilon_t). \tag{1.5}$$

Here r_t is the known, time-varying return to savings, y_t is the known part of income, and ϵ_t is the random part of income. Although it would be easy to write down the first-order condition for the maximization in the Bellman equation, the condition usually does not add much to understanding. The best approach is often just to leave the maximization to software (Matlab).

As I noted earlier, to solve for the family's optimal choice this year, x_t, start with a known value function in some distant year T, $V_T(s_T)$, iterate backwards to find V_{t+1}, and then solve the Bellman equation for the optimal x_t. Sometimes (but never in this book), the value functions have known functional forms. For the great majority of interesting problems, the functions need to be represented as well-chosen approximations. Judd (1998) discusses this topic at an advanced level. The state variables may be discrete or continuous. Discrete variables might tell whether a person was employed or unemployed, or well or sick. Discrete variables may also describe random outside events, even some, such as financial returns, that might be considered continuous. Endogenous state variables such as wealth, that result from choices that are continuous, need to be treated as continuous. One should avoid the temptation to convert them to discrete variables, as the resulting approximation is hard to manage and gives unreliable results.

Handling discrete state variables is straightforward: just subscript the value function by the discrete state. Thus the Bellman equation when s_t is discrete is

$$V_{s_t,t} = \max_{x_t}(u_t(s_t, x_t) + \beta \, \mathbb{E}_t \, V_{s_{t+1},t+1}). \qquad (1.6)$$

1.2 Approximation

Many interesting models have only a single continuous state variable, including all the models in this book. A useful family of approximations in that case is a weighted sum of known functions of s. Let $\phi_i(s)$ denote these known functions—typically there are several hundred of them. It is convenient to normalize them so that they equal 1 at a give value s_i and 0 at the points corresponding to other of the functions:

$$\phi_i(\bar{s}_i) = 1 \quad \text{and} \quad \phi_j(\bar{s}_i) = 0, \quad j \neq i. \qquad (1.7)$$

The approximation is

$$V_t(s) = \sum_i \phi_i(s) V_{i,t}. \qquad (1.8)$$

Under the normalization of the ϕs, the value $V_{i,t}$ is the value of $V_t(s)$ at $s = s_i$. The approximation interpolates between the $V_{i,t}$ points for intermediate values of s. The point defined by an s_i and V_i pair is called a *knot*.

A convenient choice for the interpolation functions is a tent:

$$\begin{aligned}
\phi_i(s) &= 0 && \text{if } s \leqslant s_{i-1} \\
&= \frac{s - s_{i-1}}{s_i - s_{i-1}} && \text{if } s_{i-1} < s \leqslant s_i \\
&= \frac{s_{i+1} - s}{s_{i+1} - s_i} && \text{if } s_i < s \leqslant s_{i+1} \\
&= 0 && \text{if } s \geqslant s_{i+1}.
\end{aligned} \tag{1.9}$$

The function $V_t(s)$ is then the linear interpolation between the knots.

Given a set of interpolation functions, the backward recursion to find the current knot values is

$$V_{i,t} = \max_{x_t}(u_t(s_i, x_t) + \beta\,\mathbb{E}_{\epsilon_t}\, U_{t+1}(f_t(s_i, x_t, \epsilon_t))). \tag{1.10}$$

For the models in this book, with twenty years of backward recursion and 500 knots in the approximation to the value function, a standard PC, vintage 2008, takes around half a minute to calculate the value functions.

When calculating the value functions, it is usually a good idea to store away the choice functions, represented as values $x_{i,t}$ of the optimal choice at time t given state variable value s_i (these are also called *policy functions*).

1.3 Stationary Case

Sometimes the stationary value function is interesting. Suppose that the decision maker is embedded in an unchanging environment with the random ϵs drawn from an unchanging distribution. Suppose further that the horizon is infinite. Then the value function becomes stationary, in the sense that it loses its time subscript. The stationary approximating Bellman equation is

$$V_i = \max_x (u(s_i, x) + \beta\, \mathbb{E}_\epsilon\, V(f(s_i, x, \epsilon))). \qquad (1.11)$$

To find the values V_i that solve this equation, we can treat the Bellman equation as a big system of nonlinear equations to be solved for the unknowns, V_i. This method is called *projection* and can be tricky, but when it works it is usually fast. An alternative, foolproof method is *value function iteration*. Start with arbitrary V_i, put them on the right-hand side of the Bellman equation, and calculate a new set. Iterate this operation to convergence, which is guaranteed (see Judd 1998, p. 412). This approach can be slow for bigger problems (Judd discusses a number of ways to speed it up).

1.4 Markov Representation

The solution to the dynamic program describes the way a decision maker responds to an uncertain environment. From the stochastic driving force ϵ, the model generates the decision maker's stochastic response, in the sense of the joint distribution of the endogenous variables of the model. Most researchers describe the joint distribution by simulation. They start at given values of the state variables s, evaluate the choice functions $x(s)$, draw a random ϵ from the appropriate distribution, compute the new state vector from the law of motion, and continue for many simulated years. But simulation is extremely inefficient: to drive down the sampling errors from simulation, which cause the joint distribution of the simulated data to differ from the true joint distribution generated by the model, to acceptable low values, you may need to simulate for days. Some (but not all) of the aspects of the joint distribution are available with essentially perfect accuracy by direct calculation rather than simulation.

A recursive model is a Markov process. For given current values of the state variables s, the choice functions and the law of motion generate a probability distribution across states in the coming year. If the model is stationary, the Markov process has constant probabilities; otherwise, they vary with time. A Markov process is fully defined by its transition matrix. Interesting aspects of the joint distribution can be calculated by standard matrix operations applied to the matrix. For example, transition probabilities over more than one year are powers of the transition matrix and the stationary probabilities of a stationary model can be calculated in no time by matrix inversion.

For a continuous state variable, the true transition matrix is infinitely big, so again we need to use an approximation. I treat the model as assuming that the state variables originate from only the grid of points used in the earlier approximation, s_i. Then I calculate the transition probability from state s_i this year to s_i' next year as the probability that a person starting from the exact point s_i this year winds up in an interval containing $s_{i'}$ next year. The interval runs from halfway between $s_{i'-1}$ and $s_{i'}$ to halfway between $s_{i'}$ and $s_{i'+1}$. I denote the transition probability as $\Pi_{i,i'}$ and calculate it as

$$\Pi_{i,i'} = \text{Prob}\left[\frac{s_{i'-1} + \bar{s}_{i'}}{2} \leqslant f(s_i, x_i, \epsilon) < \frac{s_{i'} + s_{i'+1}}{2}\right].$$
(1.12)

Solve the linear system $\pi\Pi = \pi$ and $\sum_i \pi_i = 1$ to find stationary probabilities π_i.

1.5 Distribution of the Stochastic Driving Force

The calculation of the Bellman equation requires the evaluation of an expectation over the distribution of the random ϵ. One could imagine assuming a continuous distribution of the disturbance with a known functional form and performing an analytic or numerical integration to form the expectation. But it is rare to know that the disturbance has a particular functional form and often impossible to do the integration analytically and challenging to do it numerically. It is usually better to use a purely empirical distribution. For example, if the disturbance is productivity, one can take 50 realizations of actual productivity. The integration is replaced by adding up the value function at 50 values and dividing by 50. Chapter 6 takes this approach with 14,000 values of startup companies.

2

Research on Properties of Preferences

The studies in this book use information about preferences from research on individual behavior. Consider the standard intertemporal consumption–hours problem without unemployment,

$$\max \mathbb{E}_t \sum_{\tau=0}^{\infty} \delta^{\tau} U(c_{t+\tau}, h_{t+\tau}), \tag{2.1}$$

subject to the budget constraint,

$$\sum_{\tau=0}^{\infty} R_{t,\tau}(w_{t+\tau}h_{t+\tau} - c_{t+\tau}) = 0. \tag{2.2}$$

Here $R_{t,\tau}$ is the price at time t of a unit of goods delivered at time $t + \tau$.

I let $c(\lambda, \lambda w)$ be the Frisch consumption demand and $h(\lambda, \lambda w)$ be the Frisch supply of hours per worker. See Browning et al. (1985) for a complete discussion of Frisch systems in general. The functions satisfy

$$U_c(c(\lambda, \lambda w), h(\lambda, \lambda w)) = \lambda \tag{2.3}$$

and

$$U_h(c(\lambda, \lambda w), h(\lambda, \lambda w)) = -\lambda w. \tag{2.4}$$

Here λ is the Lagrange multiplier for the budget constraint.

The Frisch functions have symmetric cross-price responses: $c_2 = -h_1$. They have three basic first-order or slope properties:

- *Intertemporal substitution in consumption,* $c_1(\lambda, \lambda w)$, the response of consumption to changes in its price.

- *Frisch labor-supply response,* $h_2(\lambda, \lambda w)$, the response of hours to changes in the wage.

- *Consumption-hours cross-effect,* $c_2(\lambda, \lambda w)$, the response of consumption to changes in the wage (and the negative of the response of hours to the consumption price). The expected property is that the cross-effect is positive, implying substitutability between consumption and hours of nonwork or complementarity between consumption and hours of work.

Consumption and hours are Frisch complements if consumption rises when the wage rises (work rises and nonwork falls); see Browning et al. (1985) for a discussion of the relation between Frisch substitution and Slutsky–Hicks substitution. People consume more when wages are high because they work more and consume less leisure. Browning et al. (1985) show that the Hessian matrix of the Frisch demand functions is negative semi-definite. Consequently, the derivatives satisfy the following constraint on the cross-effect controlling the strength of the complementarity:

$$c_2^2 \leqslant -c_1 h_2. \tag{2.5}$$

Each of these responses has generated a body of literature. In addition, in the presence of uncertainty, the curvature of U controls risk aversion, the subject of another literature.

To understand the three basic properties of consumer–worker behavior listed earlier, I draw primarily upon research at the household rather than the aggregate level. The first property is risk aversion and intertemporal substitution in consumption. With additively separable preferences across states and time periods, the coefficient of relative risk aversion and the intertemporal elasticity of substitution are reciprocals of one another. But there is no widely accepted definition of measure of substitution between pairs of commodities when there are more than two of them. Chetty (2006) discusses two natural measures of risk aversion when hours of work are also included in preferences. In one, hours are held constant, while in the other, hours adjust when the random state becomes known. He notes that risk aversion is always greater by the first measure than the second. The measures are the same when consumption and hours are neither complements nor substitutes.

The rest of this chapter summarizes the findings of recent research on the three key properties of the Frisch consumption demand and labor supply system. The own-elasticities have been studied extensively. I believe that a fair conclusion from the research is that Frisch elasticity of consumption demand is $\beta_{c,c} = -0.5$ and the Frisch elasticity of hours supply is $\beta_{h,h} = 0.7$.

The literature on measurement of the cross-elasticity is sparse, but a substantial amount of research has been done on the decline in consumption that occurs when a person moves from normal hours of work to zero because of unemployment or retirement. The ratio of unemployment consumption c_u to employment consumption c_e reflects the same properties of preferences as does the Frisch cross-elasticity. I use the parametric utility function in Hall and Milgrom (2008) to find the cross-elasticity that corresponds to the consumption ratio of 0.85. It is a Frisch cross-elasticity of $\beta_{c,h} = 0.3$.

2.1 Research Based on Marshallian and Hicksian Labor Supply Functions

The Marshallian labor supply function gives hours of work as a function of the wage and the individual's wealth. The Hicksian labor supply function replaces wealth with utility. The elasticity of the Marshallian function with respect to the wage is the uncompensated wage elasticity of labor supply and the elasticity of the Hicksian function is the compensated wage elasticity. Both are paired with consumption demand functions with the same arguments.

For simplicity, I will discuss the relation of the Marshallian and Hicksian functions to the Frisch functions used in chapter 5 with a normalization such that the elasticities are also derivatives. I consider the properties of the functions at a point normalized so that consumption, hours, the wage, and marginal utility λ are all 1. In this calibration, nonwage wealth is taken to be zero; this is not a normalization. The research I consider treats wage changes as permanent, in which case we can examine a static Marshallian labor supply function where wealth is replaced by permanent income.

From the budget constraint,

$$c(\lambda, \lambda w) - wh(\lambda, \lambda w) = x, \qquad (2.6)$$

where x is nonwage permanent income, I differentiate with respect to x, replace the derivatives of the Frisch functions with the β elasticities, and set $x = 0$ to find the Marshallian income effect:

$$-\frac{\beta_{h,h} - \beta_{c,h}}{\beta_{h,h} - \beta_{c,c} - 2\beta_{c,h}}. \tag{2.7}$$

By a similar calculation, the Marshallian uncompensated labor elasticity is

$$\beta_{h,h} - \frac{(\beta_{h,h} - \beta_{c,h})^2 + \beta_{h,h} - \beta_{c,h}}{\beta_{h,h} - \beta_{c,c} - 2\beta_{c,h}}. \tag{2.8}$$

The Hicksian compensated wage elasticity of labor supply is the difference between the Marshallian elasticity and the income effect:

$$\beta_{h,h} - \frac{(\beta_{h,h} - \beta_{c,h})^2}{\beta_{h,h} - \beta_{c,c} - 2\beta_{c,h}}. \tag{2.9}$$

The compensated elasticity is nonnegative.

Chetty (2006) takes an approach similar to the one suggested by these relations, though without explicit reference to the Frisch functions. He shows that the value of the coefficient of relative risk aversion (or, though he does not pursue the point, the inverse of the intertemporal elasticity of substitution in consumption, $-1/\beta_{c,c}$) is implied by a set of other measures. He solves for the consumption curvature parameter by drawing estimates of responses from the literature on labor supply. One is consumption–hours complementarity. The others are the compensated wage elasticity of static labor supply and the elasticity of static labor supply with respect to unearned income.

The following exercise gives results quite similar to Chetty's. From his table 1, reasonable values for the income elasticity and the compensated wage elasticity from labor supply estimates in the Marshallian–Hicksian framework are -0.11 and 0.4. For the income elasticity, the work of Imbens et al. (2001) is particularly informative. The paper tracks the response of earnings of winners of significant prizes in lotteries. It finds an income elasticity of 0.10. The range of values of the Frisch parameters that are consistent with these responses is remarkably tight with respect to the wage elasticity $\beta_{h,h}$. If the elasticity is 0.45, the complementarity parameter is $\beta_{c,h} = 0$, its minimum reasonable value, and the own-price elasticity of consumption is $\beta_{c,c} = -3.67$, an unreasonable magnitude. The minimum compensated wage elasticity is 0.4, in which case the own-price elasticity of consumption is -0.4 and the complementarity parameter is 0.4, at the outer limit of concavity. At $\beta_{h,h} = 0.402$, the other elasticities are $\beta_{c,c} = -0.53$ and $\beta_{c,h} = 0.38$, not too far from the values used in chapter 5 of $\beta_{h,h} = 0.7$, $\beta_{c,c} = -0.5$, and $\beta_{c,h} = 0.3$. The static labor-supply literature is reasonably consistent with the other research considered in this chapter. It is completely inconsistent with compensated or Frisch elasticities of labor supply in the range of 1 or above.

2.2   Risk Aversion

Research on the value of the coefficient of relative risk aversion (CRRA) falls into several broad categories. In finance, a consistent finding within the framework of the consumption capital-asset pricing model is that the CRRA has high values, in the range from 10 to 100 or more. Mehra and Prescott (1985) began this line of research. A key step in its development was Hansen and Jagannathan's (1991) demonstration that the marginal rate of substitution—the universal stochastic discounter in the consumption

CAPM—must have extreme volatility to rationalize the equity premium. Models such as that of Campbell and Cochrane (1999) generate a highly volatile marginal rate of substitution from the observed low volatility of consumption by subtracting an amount almost equal to consumption before measuring the MRS. I am skeptical about applying this approach in a model of household consumption.

A second body of research considers experimental and actual behavior in the face of small risks and generally finds high values of risk aversion. For example, Cohen and Einav (2007) find that the majority of car insurance purchasers behave as if they were essentially risk neutral in choosing the size of their deductible, but a minority are highly risk averse, so the average coefficient of relative risk aversion is about 80. But any research that examines small risks, such as having to pay the amount of the deductible or choosing among the gambles that an experimenter can offer in the laboratory, faces a basic obstacle. Because the stakes are small, almost any departure from risk neutrality, when inflated to its implication for the CRRA, implies a gigantic CRRA. The CRRA is the ratio of the percentage price discount off the actuarial value of a lottery to the percentage effect of the lottery on consumption. For example, consider a lottery with a $20 effect on wealth. At a marginal propensity to consume out of wealth of 0.05 per year and a consumption level of $20,000 per year, winning the lottery results in consumption that is 0.005% higher than losing. So if an experimental subject reports that the value of the lottery is 1%—say, 10 cents—lower than its actuarial value, the experiment concludes that the subject's CRRA is 200!

Remarkably little research has investigated the CRRA implied by choices over large risky outcomes.

One important contribution is Barsky et al. (1997). This paper finds that almost two thirds of respondents would reject a new job with a 50% chance of doubling income and a 50% chance of cutting income by 20%. The cutoff level of the CRRA corresponding to rejecting the hypothetical new job is 3.8. Only a quarter of respondents would accept other jobs corresponding to CRRAs of 2 or less. The authors conclude that most people are highly risk averse. The reliability of this kind of survey research based on hypothetical choices is an open question, though hypothetical choices have been shown to give reliable results when tied to more specific and less global choices, say, among different new products.

2.3 Intertemporal Substitution

Attanasio et al. (1999) and Attanasio and Weber (1993, 1995) are leading contributions to the literature on intertemporal substitution in consumption at the household level. These papers examine data on total consumption (not food consumption, as in some other work). They all estimate the relation between consumption growth and expected real returns from saving, using measures of returns available to ordinary households. All of these studies find that the elasticity of intertemporal substitution is around 0.7.

Barsky et al. (1997) asked a subset of their respondents about choices of the slope of consumption under different interest rates. They found evidence of quite low elasticities, around 0.2.

Guvenen (2006) tackles the conflict between the behavior of securities markets and evidence from households on intertemporal substitution. With low substitution, interest rates would be much higher than are observed. The interest rate is bounded from below by the rate of consumption growth divided by the intertemporal elasticity of substitution. Guvenen's resolution is in heterogeneity of the elasticity and highly unequal distribution of wealth. Most wealth is in the hands of those with elasticity around 1, whereas most consumption occurs among those with lower elasticity.

Finally, Carroll (2001) and Attanasio and Low (2004) have examined estimation issues in Euler equations using similar approaches. Both create data from the exact solution to the consumer's problem and then calculate the estimated intertemporal elasticity from the standard procedure, instrumental-variables estimation of the slope of the consumption-growth-interest-rate relation. Carroll's consumers face permanent differences in interest rates. When the interest rate is high relative to the rate of impatience, households accumulate more savings and are relieved of the tendency that occurs when the interest rate is lower to defer consumption for precautionary reasons. Permanent differences in interest rates result in small differences in permanent consumption growth and thus estimation of the intertemporal elasticity in Carroll's setup has a downward bias. Attanasio and Low solve a different problem, where the interest rate is a mean-reverting stochastic time series. The standard approach works reasonably well in that setting. They conclude that studies based on fairly long time-series data for the interest rate are not seriously biased. My conclusion favors studies with that character, accordingly.

I take the most reasonable value of the Frisch own-price elasticity of consumption demand to be -0.5. Again, I associate the evidence described here about the intertemporal elasticity of substitution as revealing the Frisch elasticity, even though many of the studies do not consider complementarity of consumption and hours explicitly.

2.4 Frisch Elasticity of Labor Supply

Pistaferri (2003) is a leading recent contribution to estimation of the Frisch elasticity of labor supply. This paper makes use of data on workers' personal expectations of wage change, rather than relying on econometric inferences, as has been standard in other research on intertemporal substitution. Pistaferri finds the elasticity to be 0.70 with a standard error of 0.09. This figure is somewhat higher than most earlier work in the Frisch framework or other approaches to measuring the intertemporal elasticity of substitution from the ratio of future to present wages. Here, too, I proceed on the assumption that these approaches measure the same property of preferences as a practical matter. Kimball and Shapiro (2003) survey the earlier work.

Mulligan (1998) challenges the general consensus among labor economists about the Frisch elasticity of labor supply with results showing elasticities well above 1. My discussion of the paper, published in the same volume, gives reasons to be skeptical of the finding, as it appears to flow from an implausible identifying assumption.

Kimball and Shapiro (2003) estimate the Frisch elasticity from the decline in hours of work among lottery winners, based on the assumption that the uncompensated elasticity of labor supply is 0. They find the elasticity to be about 1. But this finding is only as strong as the identifying condition.

Domeij and Floden (2006) present simulation results for standard labor supply estimation specifications suggesting that the true value of the elasticity may be double the estimated value as a result of omitting consideration of borrowing constraints.

Pistaferri studies only men and most of the rest of the literature in the Frisch framework focuses on men. Studies of labor supply generally find higher wage elasticities for women.

Rogerson and Wallenius (2009) introduce a distinction between micro and macro estimates of the Frisch elasticity, with the conclusion that the two can be quite different. Their vocabulary is different from that of chapter 5, so their conclusion does not stand in the way of the philosophy employed there, of building a macro model based on micro elasticities. By micro, they refer specifically to estimating the Frisch elasticity as the ratio of the slope of the log of hours of work over the life cycle to the slope of log-wages over the life cycle. They build a life-cycle model where most of the effects of wage variation take the form of changes in the age when people enter the labor force and when they leave, so the slope while in the labor force seriously understates the true Frisch elasticity. A nonconvex production technology is key to the understatement. Although early attempts to measure the Frisch elasticity used the approach that Rogerson and Wallenius consider, the literature I have cited here uses more robust sources of variation.

2.5   Consumption–Hours Complementarity

No research that I have found estimates the Frisch cross-elasticity directly. A substantial body of work has examined what happens to consumption when a person stops working, either because of unemployment following job loss or because of retirement, which may be the result of job loss. Under some strong assumptions, the decline in consumption identifies the cross-elasticity.

Browning and Crossley (2001) appears to be the most useful study of consumption declines during periods of unemployment. Unlike most earlier research in this area, it measures total consumption, not just food consumption. They find that the elasticity of a family member's consumption with respect to family income is 56%, for declines in income related to unemployment of that member. The actual decline in consumption upon unemployment is 14%. Low et al. (2008) confirm Browning and Crossley's finding in U.S. data from the Survey of Income Program Participation.

A larger body of research deals with the "retirement consumption puzzle," the decline in consumption thought to occur upon retirement. Most of this research considers food consumption. Aguiar and Hurst (2005) show that, upon retirement, people spend more time preparing food at home. The change in food consumption is thus not a reasonable guide to the change in total consumption. Hurst (2008) surveys this research.

Banks et al. (1998) use a large British survey of annual cross sections to study the relation between retirement and consumption of nondurables. They compare annual consumption changes in four-year-wide cohorts, finding a coefficient of -0.26 on a dummy for households where the head left the labor market between the two surveys. They use earlier data as instruments, so they interpret the finding as measuring the planned reduction in consumption upon retirement.

Miniaci et al. (2003) fit a detailed model to Italian cohort data on nondurable consumption, in a specification of the level of consumption that distinguishes age effects from retirement effects. The latter are broken down by age of the household head. The pure retirement reductions range from 4 to 20%. This study also finds pure unemployment reductions in the range discussed above.

Fisher et al. (2005) study total consumption changes in the Consumer Expenditure Survey, using cohort analysis. They find small declines in total consumption associated with rising retirement among the members of a cohort. Because retirement in a cohort is a gradual process and because retirement effects are combined with time effects on a cohort analysis, it is difficult to pin down the effect.

In the parametric preferences considered in Hall and Milgrom (2008), a difference in consumption between workers and nonworkers of 15% corresponds to a Frisch cross-price elasticity of demand of 0.3, the value I adopt.

3
Health

3.1 The Issues

In the American health system, families make some of
the important decisions about health spending. They
do so directly by some choices about when to seek
care, and more indirectly as voices at their employers,
who make choices about insurance coverage, and as
citizens, because the government has a large role in
financing health care for people over 64. In this chapter, based on a joint paper with my colleague Charles
Jones (Hall and Jones 2007), a family dynamic program governs the choice between current consumption and investment in health. The model helps explain
the growth of GDP spent on health.

Over the past half century, Americans spent a rising share of total economic resources on health and
enjoyed substantially longer lives as a result. Debate
on health policy often focuses on limiting the growth
of health spending. Is the growth of health spending a
rational response to changing economic conditions—
notably the growth of income per person? A dynamic-
programming model of family choice about health
spending and consumption suggests that this is indeed the case. Standard preferences—of the kind used

widely in economics to study consumption, asset pricing, and labor supply—imply that health spending is a superior good with an income elasticity well above 1. As people get richer and consumption rises, the marginal utility of consumption falls rapidly. Spending on health to extend life allows individuals to purchase additional periods of utility. The marginal utility of life extension does not decline. As a result, the optimal composition of total spending shifts toward health, and the health share grows along with income. In projections based on the quantitative analysis of our model, the optimal health share of spending seems likely to exceed 30% by the middle of the century.

The share of health care in total spending was 5.2% in 1950, 9.4% in 1975, and 15.4% in 2000. Over the same period, health has improved. The life expectancy of an American born in 1950 was 68.2 years, of one born in 1975, 72.6 years, and of one born in 2000, 76.9 years.

Why has this health share been rising, and what is the likely time path of the health share for the rest of the century? In the model, the key decision is the division of total resources between health care and non-health consumption. Utility depends on quantity of life—life expectancy—and quality of life—consumption. People value health spending because it allows them to live longer and to enjoy better lives.

Standard preferences imply that health is a superior good with an income elasticity well above 1. As people grow richer, consumption rises but they devote an increasing share of resources to health care. The quantitative analysis suggests these effects can be large: projections in our model typically lead to health shares that exceed 30% of GDP by the middle of this century.

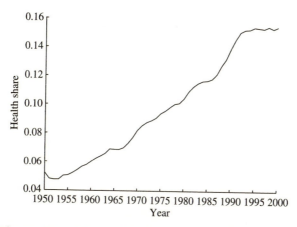

Figure 3.1. The health share in the United States.

The model abstracts from the complicated institutions that shape spending in the United States and asks a more basic question: from a social welfare standpoint, how much should the nation spend on health care, and what is the time path of optimal health spending?

3.2 Basic Facts

The appropriate measure of total resources is consumption plus government purchases of goods and services. Investment and net imports are intermediate products. Figure 3.1 shows the fraction of total spending devoted to health care, according to the U.S. National Income and Product Accounts. The numerator is consumption of health services plus government purchases of health services, and the denominator is consumption plus total government purchases of goods and services. The fraction has a sharp upward trend, but growth is irregular. In particular, the

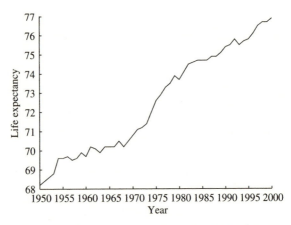

Figure 3.2.   Life expectancy in the United States.

fraction grew rapidly in the early 1990s and flattened in the late 1990s. Not shown in the figure is the resumption of growth after 2000.

Figure 3.2 shows life expectancy at birth for the United States. Following the tradition in demography, this life expectancy measure is not expected remaining years of life (which depends on unknown future mortality rates), but is life expectancy for a hypothetical individual who faces the cross section of mortality rates from a given year. Life expectancy has grown about 1.7 years per decade. It shows no sign of slowing over the fifty years reported in the figure. In the first half of the twentieth century, however, life expectancy grew at about twice this rate, so a longer times series would show some curvature.

3.3   Basic Model

The basic model makes the simple but unrealistic assumption that mortality is the same in all age groups.

Preferences are unchanging over time, and income and productivity are constant. This model sets the stage for the full model incorporating age-specific mortality and productivity growth.

The economy contains people of different ages who are otherwise identical, grouped in families who make the economic decisions. All families and all family members are identical. Let x denote a person's *health status*. The mortality rate of an individual is the inverse of her health status, $1/x$. Because people of all ages face this same mortality rate, x is also equal to life expectancy. For simplicity at this stage, time preference is zero.

In the stationary environment, the family's value function is a constant V times the number of members of the family N, which is the only state variable. The family's dynamic program is

$$VN_t = \max(n_t u(c) + VN_{t+1}). \tag{3.1}$$

The law of motion of family size is

$$N_{t+1} = \left(1 - \frac{1}{x}\right)N_t. \tag{3.2}$$

Putting the law of motion into the dynamic program and solving for V yields expected lifetime utility per family member:

$$V = xu(c). \tag{3.3}$$

In this stationary environment, consumption is constant so that expected utility is the number of years an individual expects to live multiplied by per-period utility. Period utility depends only on consumption; see Hall and Jones (2007) for a discussion of the extension to making the quality of life depend on health status. Here and throughout the paper, utility after death is zero.

Rosen (1988) pointed out the following important implication of a specification of utility involving life expectancy. When lifetime utility is per-period utility, u, multiplied by life expectancy, the level of u matters a great deal. In many other settings, adding a constant to u has no effect on consumer choice. Here, adding a constant raises the value the consumer places on longevity relative to consumption of goods. Negative utility also creates an anomaly—indifference curves have the wrong curvature and the first-order conditions do not maximize utility. As long as u is positive, preferences are well behaved.

The representative family member receives a constant flow of resources y that can be spent on consumption or health:

$$c + h = y. \tag{3.4}$$

The economy has no physical capital or foreign trade that permits shifting resources from one period to another.

Finally, a health production function governs the individual's state of health:

$$x = f(h). \tag{3.5}$$

The family chooses consumption and health spending to maximize the utility of the individual in equation (3.3) subject to the resource constraint of equation (3.4) and the production function for health status equation (3.5). That is, the optimal allocation solves

$$\max_{c,h} f(h)u(c) \quad \text{s.t. } c + h = y. \tag{3.6}$$

The optimal allocation equates the ratio of health spending to consumption to the ratio of the elasticities of the health production function and the flow utility function. With $s \equiv h/y$, the optimum is

$$\frac{s^*}{1 - s^*} = \frac{h^*}{c^*} = \frac{\eta_h}{\eta_c}, \tag{3.7}$$

where

$$\eta_h \equiv f'(h)\frac{h}{x} \quad \text{and} \quad \eta_c \equiv u'(c)\frac{c}{u}.$$

Now ignore the fact that income and life expectancy are taken as constant in this static model and instead consider what happens if income grows. The short-cut of using a static model to answer a dynamic question anticipates the findings of the full dynamic model quite well.

The response of the health share to rising income depends on the movements of the two elasticities in equation (3.7). The crux of the argument is that the consumption elasticity falls relative to the health elasticity as income rises, causing the health share to rise. Health is a superior good because satiation occurs more rapidly in nonhealth consumption.

Why is η_c decreasing in consumption? In most branches of applied economics, only marginal utility matters. For questions of life and death, however, this is not the case. The utility associated with death is normalized at zero in our framework, and how much a person will pay to live an extra year hinges on the level of utility associated with life. In our application, adding a constant to the flow of utility $u(c)$ has a material effect: it permits the elasticity of utility to vary with consumption. Thus the approach takes the standard constant-elastic specification for marginal utility but adds a constant to the level of utility.

What matters for the choice of health spending, however, is not just the elasticity of marginal utility, but also the elasticity of the flow utility function itself. With the constant term added to a utility function with constant-elastic marginal utility, the utility elasticity declines with consumption for conventional parameter values. The resulting specification is then capable of explaining the rising share of health spending.

Period utility is

$$u(c) = b + \frac{c^{1-y}}{1-y}, \tag{3.8}$$

where y is the constant elasticity of marginal utility. Based on the evidence in chapter 2, $y > 1$ is the likely case. Accordingly, the second term is negative, so the base level of utility, b, needs to be positive enough to ensure that flow utility is positive over the relevant values of c. The flow of utility $u(c)$ is always less than b, so the elasticity η_c is decreasing in consumption. More generally, any bounded utility function $u(c)$ will deliver a declining elasticity, at least eventually, as will the unbounded $u(c) = \alpha + \beta \log c$. Thus the key to the explanation of the rising health share—a marginal utility of consumption that falls sufficiently quickly— is obtained by adding a constant to a standard class of utility functions.

A rapidly declining marginal utility of consumption leads to a rising health share provided the health production elasticity η_h does not itself fall too rapidly. For example, if the marginal product of health spending in extending life were to fall to zero—say, it was technologically impossible to live beyond the age of 100—then health spending would cease to rise at that point. Whether or not the health share rises over time is then an empirical question: there is a race between diminishing marginal utility of consumption and the diminishing returns to the production of health. As

discussed later, for the kind of health production functions that match the data, the production elasticity declines very gradually, and the declining marginal utility of consumption does indeed dominate, producing
a rising health share.

The simple model develops intuition, but it falls
short on a number of dimensions. Most importantly,
the model assumes constant total resources and constant health productivity. This means it is inappropri-
ate to use this model to study how a growing income
leads to a rising health share, the comparative static
results notwithstanding. Still, the basic intuition for a
rising health share emerges clearly. The health share
rises over time as income grows if the marginal utility of consumption falls sufficiently rapidly relative to
the joy of living an extra year and the ability of health
spending to generate that extra year.

3.4   The Full Dynamic-Programming Model

The full dynamic-programming model has age-specific
mortality, growth in total resources, and productivity
growth in the health sector. All families have the same
age composition.

An individual of age a in period t has an age-specific
state of health, $x_{a,t}$. As in the basic model, the mortality rate for an individual is the inverse of her health
status. Therefore, $1 - 1/x_{a,t}$ is the per-period survival
probability of an individual with health $x_{a,t}$.

An individual's state of health is produced by spending on health $h_{a,t}$:

$$x_{a,t} = f(h_{a,t}; a, t). \tag{3.9}$$

In this production function, health status depends on
both age and time. Forces outside the model that vary

with age and time may also influence health status—examples include technological change and education.
The flow utility of the individual is

$$u(c_{a,t}, x_{a,t}) = b + \frac{c_{a,t}^{1-\gamma}}{1-\gamma}. \tag{3.10}$$

The first term is the baseline level of utility discussed earlier. The second is the standard constant-elastic specification for consumption.

In this environment, consider the allocation of resources that would be chosen by a family who places equal weights on each person alive at a point in time and who discounts future flows of utility at rate β. Let $N_{a,t}$ denote the number of people of age a alive at time t. Let $V_t(N_t)$ denote the family's value function when the age distribution of the population is the vector $N_t \equiv (N_{1,t}, N_{2,t}, \dots, N_{a,t}, \dots)$. Then the family's dynamic program is

$$V_t(N_t)$$

$$= \max_{\{h_{a,t}, c_{a,t}\}} \left(\sum_{a=0}^{\infty} N_{a,t}\, u(c_{a,t}, x_{a,t}) + \beta V_{t+1}(N_{t+1}) \right) \tag{3.11}$$

with law of motion

$$N_{a+1,t+1} = \left(1 - \frac{1}{x_{a,t}}\right) N_{a,t}, \tag{3.12}$$

subject to

$$\sum_{a=0}^{\infty} N_{a,t}(y_t - c_{a,t} - h_{a,t}) = 0, \tag{3.13}$$

$$N_{0,t} = N_0, \tag{3.14}$$

$$x_{a,t} = f(h_{a,t}; a, t), \tag{3.15}$$

$$y_{t+1} = e^{g_y} y_t. \tag{3.16}$$

The law of motion for the population assumes a large enough family so that the number of people aged $a + 1$ next period can be taken equal to the number aged a today multiplied by the survival probability. This assumption is plainly unrealistic as far as family composition. The assumption eliminates the need to keep track of the distribution of age compositions across families. General-equilibrium models often make this kind of assumption for tractability.

The first constraint is the family-wide resource constraint. Note that people of all ages contribute the same flow of resources, y_t. The second constraint specifies that births are exogenous and constant at N_0. The final two constraints are the production function for health and the law of motion for resources, which grow exogenously at rate g_y.

Let λ_t denote the Lagrange multiplier on the resource constraint. The optimal allocation satisfies the following first-order conditions for all a:

$$u_c(c_{a,t}) = \lambda_t, \quad (3.17)$$

$$\beta \frac{\partial V_{t+1}}{\partial N_{a+1,t+1}} \frac{f'(h_{a,t})}{x_{a,t}^2} + u_x(c_{a,t}) f'(h_{a,t}) = \lambda_t, \quad (3.18)$$

where $f'(h_{a,t})$ represents $\partial f(h_{a,t}; a, t)/\partial h_{a,t}$. That is, the marginal utility of consumption and the marginal utility of health spending are equated across family members and to each other at all times. This condition together with the additive separability of flow utility implies that members of all ages have the same consumption c_t at each point in time, but they have different health expenditures $h_{a,t}$ depending on age.

Normally, a value function with many state variables creates substantial computational challenges. In this model, none arise because the value function is known to be linear:

$$V_t(N_t) = \sum_a v_{a,t} N_{a,t}; \qquad (3.19)$$

$v_{a,t}$ is the value of life at age a in units of utility. Combining the two first-order conditions yields

$$\frac{\beta v_{a+1,t+1}}{u_c} + \frac{u_x x_{a,t}^2}{u_c} = \frac{x_{a,t}^2}{f'(h_{a,t})}. \qquad (3.20)$$

The optimal allocation sets health spending at each age to equate the marginal benefit of saving a life to its marginal cost. The marginal benefit is the sum of two terms. The first is the social value of life $\beta v_{a+1,t+1}/u_c$. The second is the additional quality of life enjoyed by people as a result of the increase in health status.

Taking the derivative of the value function shows that the social value of life satisfies the recursive equation:

$$v_{a,t} = u(c_t) + \beta\left(1 - \frac{1}{x_{a,t}}\right) v_{a+1,t+1}$$
$$+ \lambda_t(y_t - c_t - h_{a,t}). \qquad (3.21)$$

The additional social welfare associated with having an extra person alive at age a is the sum of three terms. The first is the level of flow utility enjoyed by that person. The second is the expected social welfare associated with having a person of age $a + 1$ alive next period, where the expectation employs the survival probability $1 - 1/x_{a,t}$. Finally, the last term is the net social resource contribution from a person of age a, her production less her consumption and health spending.

The literature on competing risks of mortality suggests that a decline in mortality from one cause may increase the optimal level of spending on other causes, as discussed by Dow et al. (1999). This property holds in this model as well. Declines in future mortality will increase the value of life, $v_{a,t}$, raising the marginal benefit of health spending at age a.

3.5   The Health Production Function

A period in the model is five years. The data are in twenty five-year age groups, starting at 0–4 and ending at 95–99. The historical period has eleven time periods in running from 1950 through 2000. See Hall and Jones (2007) for a discussion of data sources.

The inverse of the nonaccident mortality rate (adjusted health status), $\tilde{x}_{a,t}$ is a Cobb–Douglas function of health inputs:

$$\tilde{x}_{a,t} = A_a (z_t h_{a,t} w_{a,t})^{\theta_a}. \tag{3.22}$$

In this production function, A_a and θ_a are parameters that depend on age. z_t is the efficiency of a unit of output devoted to health care, taken as an exogenous trend; it is the additional improvement in the productivity of health care on top of the general trend in the productivity of goods production. The unobserved variable $w_{a,t}$ captures the effect of all other determinants of mortality, including education and pollution.

Hall and Jones (2007) discusses estimation of the production parameters. Figure 3.3 shows the estimates of θ_a, the elasticity of adjusted health status, $\tilde{x}$, with respect to health inputs, by age category. The groups with the largest improvements in health status over the fifty-year period, the very young and the middle-aged, have the highest elasticities, ranging from 0.25 to 0.40. The fact that the estimates of

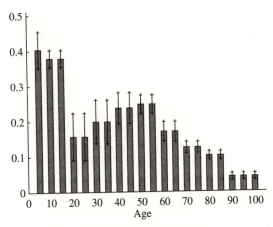

Figure 3.3. Estimated effects of
health inputs on health status.

θ_a generally decline with age, particularly at the older
ages, constitutes an additional source of diminishing
returns to health spending as life expectancy rises. For
the oldest age groups, the elasticity of health status
with respect to health inputs is only 0.042.

3.6 Preference Parameters

Chapter 2 discussed evidence on the curvature parameter of the utility function, y, with the conclusion that
$y = 2$ is the most reasonable value. Alternative values range from near-log utility ($y = 1.01$) to $y = 2.5$.
The discount factor, β, is consistent with the choice
of y and a 6% real return to saving. With consumption
growth from the data of 2.08% per year, a standard Euler equation gives an annual discount factor of 0.983,
or, for the five-year intervals in the model, 0.918.

The intercept of flow utility b is determined to imply
a $3 million value of life for 35–39-year-olds in the year

2000. Hall and Jones (2007) discusses this calculation in detail, including many references to research on the value of life.

3.7 Solving the Model

For the historical period 1950–2000, the solution uses resources per person, y, at its actual value. For the projections into the future, it assumes income continues to grow at its average historical rate of 2.31% per year. Solution is by value function iteration, starting with arbitrary values of the coefficients of the value function $v_{a,T}$ for a distant horizon T, solving equation (3.20) successively for each earlier period, evaluating the v coefficients for the next earlier period, and repeating back to 1950.

Figure 3.4 shows the calculated share of health spending over the period 1950 through 2050 for y between 1.01 and 2.5. A rising health share is a robust feature of the optimal allocation of resources in the health model, as long as y is not too small. The curvature of marginal utility, y, is a key determinant of the slope of optimal health spending over time. If marginal utility declines quickly because y is high, the optimal health share rises rapidly. This growth in health spending reflects a value of life that grows faster than income. In fact, in the simple model, the value of a year of life is roughly proportional to c^y, illustrating the role of y in governing the slope of the optimal health share over time.

For near-log utility (where $y = 1.01$), the optimal health share declines. The reason for this is the declining elasticity of health status with respect to health spending in our estimated health production technology (recall figure 3.3). In this case, the marginal utility

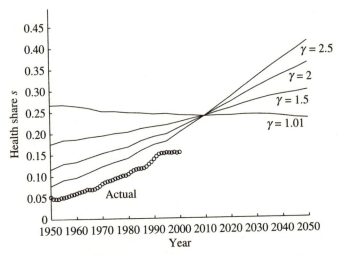

Figure 3.4. Health share of spending from the model for various values of the curvature parameter γ.

of consumption falls sufficiently slowly relative to the diminishing returns in the production of health that the optimal health share declines gradually over time. The careful reader might wonder why all of the optimal health shares intersect in the same year, around 2010. This intersection follows from choosing the utility intercept b to match a specific level for the value of life for 35–39-year-olds and to the fact that our preferences feature a constant elasticity of marginal utility.

3.8 Concluding Remarks

A model based on standard economic assumptions yields a strong prediction for the health share. Provided the marginal utility of consumption falls sufficiently rapidly—as it does for an intertemporal elasticity of substitution well under 1—the optimal health share rises over time. The rising health share occurs as consumption continues to rise, but consumption grows more slowly than income. The intuition for this result is that in any given period, people become saturated in nonhealth consumption, driving its marginal utility to low levels. As people get richer, the most valuable channel for spending is to purchase additional years of life. The numerical results suggest the empirical relevance of this channel: optimal health spending is predicted to rise to more than 30% of GDP by the year 2050 in most of our simulations, compared with the current level of about 15%.

This fundamental mechanism in the model is supported empirically in a number of different ways. First, as discussed earlier, it is consistent with conventional estimates of the intertemporal elasticity of substitution. Second, the mechanism predicts that the value of a statistical life should rise faster than income. This is a strong prediction of the model, and a place where careful empirical work in the future may be able to shed light on its validity. Costa and Kahn (2004) and Hammitt et al. (2000) provide support for this prediction, suggesting that the value of life grows roughly twice as fast as income, consistent with our baseline choice of $\gamma = 2$. Cross-country evidence also suggests that health spending rises more than one-for-one with income; this evidence is summarized by Gerdtham and Jonsson (2000).

One source of evidence that runs counter to our prediction is the micro evidence on health spending and income. At the individual level within the United States, for example, income elasticities appear to be substantially less than 1, as discussed by Newhouse (1992). A serious problem with this existing evidence, however, is that health insurance limits the choices facing individuals, potentially explaining the absence of income effects. Future empirical work will be needed to judge this prediction.

As mentioned in the introduction, the recent health literature has emphasized the importance of technological change as an explanation for the rising health share. In the view developed here, technology change is a proximate rather than a fundamental explanation. The development of new and expensive medical technologies is surely part of the process of rising health spending, as the literature suggests; Jones (2004) provides a model along these lines with exogenous technical change. However, a more fundamental analysis looks at the reasons that new technologies are developed. Distortions associated with health insurance in the United States are probably part of the answer, as suggested by Weisbrod (1991). But the fact that the health share is rising in virtually every advanced country in the world—despite wide variation in systems for allocating health care—suggests that deeper forces are at work. A fully worked-out technological story will need an analysis on the preference side to explain why it is useful to invent and use new and expensive medical technologies. The most obvious explanation is the model here: new and expensive technologies are valued because of the rising value of life.

Viewed from every angle, the results support the proposition that both historical and future increases in the health spending share are desirable. The magnitude of the future increase depends on parameters whose values are known with relatively low precision, including the value of life, the curvature of marginal utility, and the fraction of the decline in age-specific mortality that is due to technical change and the increased allocation of resources to health care. Nevertheless, it seems likely that maximizing social welfare in the United States will require the development of institutions that are consistent with spending 30% or more of GDP on health by the middle of the century.

4

Insurance

The study of insurance fits naturally into a dynamic program. The family's value function is almost always concave in wealth, so the family will want to trade a small payment made with certainty—the insurance premium—to avoid a large loss in wealth from an insurable event, such as a car accident, fire, or disability.

This chapter looks at one important type of insurance, for long-term care, through the lens of a study by Brown and Finkelstein (2008) that exemplifies the modern approach to the analysis of policy questions. Long-term care in nursing homes and similar facilities accounts for 1.2% of GDP and almost 9% of health spending. But only 10% of likely customers—family members over 65—purchase private insurance for long-term care. Most people go without insurance and pay for care as they receive it, or take advantage of Medicaid, the public program that pays for care for families whose savings are exhausted and whose incomes are below a threshold (the social security program for people over 64, Medicare, does not pay for long-term care). The policy question is whether the availability of Medicaid discourages the purchase of private insurance. The answer is emphatically yes. The authors' dynamic program model sheds important light on how Medicaid could be redesigned to improve the efficiency of the allocation of long-term care, by bringing a larger fraction of the population under insurance. A big problem currently is that the limited insurance under Medicaid is enough to discourage people from buying private insurance, even though they are not always eligible for Medicaid support.

4.1 The Model

People in the model are in one of five states, indexed by a discrete state variable s:

  (i) no care needed;
 (ii) receiving care at home;
(iii) receiving care in an assisted-living facility;
(iv) receiving care in a nursing home;
 (v) dead.

The use of nonhome care is fairly high—12% of men and 20% of women are in assisted-living facilities at some time after age 65, and 27% of men and 44% of women are in nursing homes at some time. Because married women substantially outlive their husbands on average, men are often cared for at home by their wives, who then enter nursing homes after their husbands die.

Although the paper is firmly rooted in a dynamic program, you need to download the appendix from the *AER* website to see it written out (Google "AER Brown Finkelstein").

The individual's period utility function has constant elasticity:

$$U(C + F) = \frac{(C + F)^{1-\gamma}}{1 - \gamma}. \qquad (4.1)$$

People consume in two ways, as an amount chosen and paid for directly by the individual, C, and an amount provided by a facility, F_s, for $s = 3$ and $s = 4$; $F_1 = F_2 = 0$.

A person has a continuous state variable, assets, W. The law of motion for wealth contains all the special institutional aspects of long-term care and Medicaid. The ingredients are A, the individual's annuity retirement income, B, the maximum benefit that private insurance pays, X, out-of-pocket long-term care expenditures by the individual, P, the private insurance premium, and r, the monthly real rate of interest.

The first issue is eligibility for Medicaid. An eligible person needs care ($s = 2$, 3, or 4), has assets below a threshold ($W < \underline{W}$), and has income $A + rW$ below a threshold $\underline{C}$ plus expenditures on care not reimbursed by private insurance. Although eligibility depends on past choices, given the state variables, it is determined as of the beginning of the period.

For an individual not eligible for Medicaid, the law of motion for assets is

$$W_{t+1} = (1+r)(W_t + A_t + \min(B_{s,t}, X_{s,t}) - P_{s,t} - C) \quad (4.2)$$

and for those on Medicaid

$$W_{t+1} = (1 + r)(\min(W_t, \underline{W}) + \underline{C} - C). \quad (4.3)$$

The individual's dynamic program is

$$V_{s,t}(W_t) = \max_C \left(U(C + F_{s,t}) + \frac{1}{1 + \rho} \sum_{s'} q_t^{s,s'} V(W_{t+1}) \right). \quad (4.4)$$

Here ρ is the rate of impatience and q is the transition matrix from state s to state s'.

The authors solve the dynamic program by backward value-function recursion, starting at age 105, using a grid of 1,400 values of W.

Table 4.1. Annual transition matrix among health-care states.

	State next year				
State this year	No care	Home care	Assisted living	Nursing home	Dead
No care	91	2.1	0.4	1.9	4.4
Home care	12	63	1.2	10	14
Assisted living	14	41	19	15	11
Nursing home	7.4	19	7.7	48	17
Dead	0	0	0	0	100

4.2 Calibration

Brown and Finkelstein take the coefficient of relative risk aversion as $y = 3$, higher than the value advocated in chapter 2, but they report that the conclusions also hold with less risk aversion.

The authors use a transition matrix that is an approximation of a well-established model of the dynamics of health care. The model has a number of state variables and complicated patterns of time dependence. The approximation is a monthly time-varying Markov process. Table 4.1 shows the 12th power of the transition matrix for eighty-year-old women. Those receiving no care are 91% likely to require no care in the following year and only 4.4% likely to die. Those receiving home care have a 12% likelihood of recovering and requiring no care and about a 25% likelihood of moving on to more intensive care or dying. Assisted living is by far the least persistent state—more than half in that state will improve to no care or home care in the following year and about a quarter will move to nursing homes or die. The mortality rate in nursing homes is 17% per year.

Costs for care are $2,160 per month for assisted living and $4,290 for nursing homes. They do not report the cost of home care but state that it is substantially less expensive. The Medicaid ceiling on assets, $\underline{W}$, is $2,000 and the ceiling on income, $\underline{C}$, is $30 per month for those in assisted living and nursing homes and $545 per month for home care. The value of consumption supplied in facilities is $515 per month.

4.3   Results

The authors state the results of the analysis in terms of the willingness to pay for insurance. They find the value function $\tilde{V}$ for an individual in the environment of the model, except that private insurance is not available. They then solve the equation,

$$V(W) = \tilde{V}(W + P(W)), \qquad (4.5)$$

for the extra wealth, $P(W)$, needed to compensate the individual denied access to insurance so she achieves the same level of expected utility as she would with insurance. Willingness to pay depends on the reference level of wealth, W. Chapter 6 shows how to build this calculation directly into the dynamic program, rather than doing it later, but the results are the same.

Figure 4.1 shows the willingness-to-pay function $P(W)$ for men and women. The horizontal axis is the location of the individual in the relevant wealth distribution, specified as a percentile. The median (50th percentile) wealth is about $220,000. About 60% of this wealth takes the form of the present value of retirement annuities. The private insurance pays up to

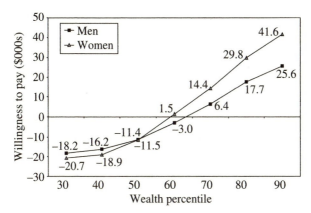

Figure 4.1. Willingness to pay for
private long-term care insurance.

$3,000 per month and its price involves a *load*, the difference between the price and the actual present value of the benefit. Men and women pay the same price, but the insurance is way more valuable to women than to men, for the reason mentioned before—men are much more likely to have a wife to provide care at home than women are to have a husband. The load for men is 50%—the benefits are only half the price—and for women, −6%. Equal pricing for men and women generates a large cross-subsidy, because the cost is so much higher for women. Brown and Finkelstein do not take up the policy issue of whether equal pricing has social benefits over pricing closer to cost.

For both men and women, willingness to pay rises with wealth. The slope is the reverse of what would be found for most kinds of insurance. Chapter 6 will show that insurance against the idiosyncratic risk of entrepreneurship is more valuable for those with *less* wealth. The unusual upward slope arises from the role of Medicaid—people with less wealth are much more likely to deplete their wealth down to the low level that makes them eligible for Medicaid coverage of long-term care. People with, say, half a million dollars of wealth (at the 80th percentile) are unlikely to pursue that strategy, so, lacking insurance from Medicaid, they value private insurance.

For both men and women, the cutoff level of wealth is a little below $300,000 (60th percentile). Above this level, the model says that people should buy long-term care insurance. Below the cutoff, Medicaid provides enough insurance that private insurance is not worth the cost.

If loads were zero for private insurance separately priced for men, the situation would not be terribly different (loads are close to zero for women already). The cutoff wealth level would fall to $220,000 (50th percentile). Medicaid would continue to crowd out private insurance for half of the men.

The role of Medicaid in discouraging the purchase of private long-term care insurance can be expressed in the form of an implicit tax on private insurance. For men the tax rate is 99.8% in the lowest wealth decile, 59% at the median wealth, and 3% in the highest decile. Implicit tax rates are even higher for women, at 99.9%, 77%, and 5%.

Medicaid prohibits people from buying insurance that coordinates with Medicaid benefits by covering only the cost of long-term care that Medicaid does not cover. The authors calculate the lost value from this prohibition as the willingness to pay for an actuarially fair (zero load) insurance policy of the prohibited type. The lost value is zero for those in the bottom 20% of the wealth distribution, is $20,000 for men and $30,000 for women at median wealth, and is $101,000 for men and $166,000 for women in the top 10% of the wealth distribution.

The constant-elastic (constant relative risk aversion) utility function in the Brown–Finkelstein model has the property that people are enormously concerned about even small probabilities of low consumption—the marginal utility of consumption approaches infinity as consumption approaches zero. They investigate the possible role of this property in two ways: by adopting a constant absolute risk aversion (exponential) utility function and by positing that relatives and charity place a floor on consumption. Both alternative specifications replicate the main finding that Medicaid substantially displaces private insurance. They also make a general argument that the displacement invariably implies that the total amount of insurance is less than would be purchased if Medicaid did not exist but the public could buy actuarially fair zero-load private insurance. The reason is that even the wealthiest risk-averse people will buy such insurance.

The authors caution that all of these conclusions flow from a model with strong assumptions about rational economic behavior. Not everybody buys insurance when it is reasonably priced. The model omits factors such as adverse selection that are known to impede insurance markets. But the paper is a leading example of what can be learned from a family dynamic program that makes a serious attempt to build in features of the economy that matter for policy making.

5

Employment

Dynamic choices of families are central to aggregate movements of key variables: hours of work, the employment rate, and consumption. This chapter asks how far the received principles of family choice can take the economist in understanding the cyclical properties of these variables. Figure 5.1 shows the data whose joint movements I seek to understand. The data are detrended to focus on the cyclical movements. The series are nondurables and services consumption per person, weekly hours per worker, the employment rate (fraction of the labor force working in a given week, 1 minus the unemployment rate), and the average product of labor for the United States. Common movements associated with the business cycle are prominent in all four measures. Consumption and productivity are fairly well correlated with each other and so are hours and employment. The correlation of productivity with hours and employment is lower, especially in the last fifteen years of the sample.

The model in this paper considers a worker in a representative family that maximizes the expected discounted sum of future utility. Unlike the models in the other chapters, I do not actually solve a dynamic program. Instead I ask if behavior is consistent with time-separable preferences with the elasticities discussed

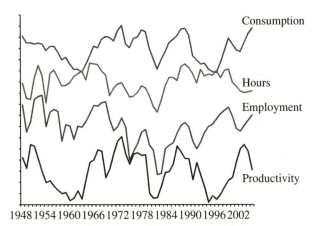

1948 1954 1960 1966 1972 1978 1984 1990 1996 2002

Figure 5.1. Detrended consumption, hours per worker, employment rate, and productivity. The tick marks on the vertical axis are one percentage point apart. Constants are added to the series to separate them vertically.

in chapter 2. The family chooses consumption and the hours of workers according to completely standard principles. I find the Frisch system of consumption demand and weekly hours supply equations the most convenient way to embody these principles.

I do not assume that the members of the family are on their labor supply functions. Frictions in the labor market prevent market clearing in that sense. The family takes friction in the labor market as given—though the family would allocate all of its eligible members to working if it could, in fact only a fraction of them are working at any moment because the remainder are searching for work. A key element of the model is a function that relates the employment rate to the same variables that control the consumption and hours choices of the family. Even though the family takes the employment rate as a given feature of the labor market, the employment rate resulting from the inter-action of all workers and firms depends on marginal utility of goods consumption and the marginal product of labor. Thus the model describes consumption, hours, and the employment rate in terms of just two factors.

In this chapter, I do not consider the small procyclical movements of participation in the labor force—Hall (2008) documents these movements. For simplicity, I treat the labor force as exogenously determined.

I treat marginal utility and the common value of the marginal value of time and the marginal product of labor as unobserved latent state variables. I take each of the four indicators—consumption, hours, the employment rate, and productivity—as a function of the two latent variables plus an idiosyncratic residual. I do not use macro data to estimate the model's slope parameters. One reason is that the model falls short of identification. The main reason is that macro data are probably not the best way to estimate parameters; data at the household level are generally more powerful. I use information from extensive research on some of the coefficients.

5.1   Insurance

By adopting the same device as in chapter 3—an assumption that people live in such large families that the law of large numbers applies to them—the analysis in this chapter makes the assumption that workers are insured against the personal risk of the labor market and that the insurance is actuarially fair. The insurance makes payments based on outcomes outside the control of the worker that keep all workers' marginal

utility of consumption the same. This assumption—
dating at least back to Merz (1995)—results in enormous analytical simplification. In particular, it makes
the Frisch system of consumption demand and labor
supply the ideal analytical framework. Absent the assumption, the model is an approximation based on
aggregating employed and unemployed individuals,
each with a personal state variable, wealth. Blundell
et al. (2008) find evidence of substantial insurance of
individual workers against transitory shocks such as
unemployment.

I do not believe that, in the U.S. economy, consumption during unemployment behaves literally according to the model with full insurance against unemployment risk. But families and friends may provide
partial insurance. I view the fully insured case as a
good and convenient approximation to the more complicated reality, where workers use savings and partial insurance to keep consumption close to the levels
that would maintain roughly constant marginal utility. I make no claim that workers are insured against
idiosyncratic permanent changes in their earnings ca-
pacities, only that the transitory effects of unemployment can usefully be analyzed under the assumption
of insurance.

5.2 Dynamic Labor-Market Equilibrium

This section develops a unified model of the labor
market and production. The outcome is a set of four
equations relating the four observed variables in figure 5.1—consumption, weekly hours, the employment
rate, and the tax-adjusted marginal product of labor—
to a pair of latent variables: the marginal utility of

consumption and the marginal value of time. For consumption and hours, I use the Frisch consumption demand and hours supply equations, which provide a direct connection to a large body of research on household behavior in the Frisch framework. For the employment rate, I use an equation relating the employment rate to marginal utility and the marginal value of time derived from a generalization of the search and matching model.

I consider an economy with many identical families, each with a large number of members. All workers face the same pay schedule and all members of all families have the same preferences. The family insures its members against personal (but not aggregate) risks and satisfies the Borch–Arrow condition for optimal insurance of equal marginal utility across individuals. In each family, a fraction n_t of workers are employed and the remaining $1 - n_t$ are searching. These fractions are outside the control of the family—they are features of the labor market. In my calibration, a family never allocates any of its members to pure leisure—it achieves higher family welfare by assigning all nonworking members to job search and it never terminates the work of an employed member. Thus, as I noted earlier, I neglect the small variations in labor-force participation that occur in the actual U.S. economy. To generate realistically small movements of participation in the model I would need to introduce heterogeneity in preferences or earning powers.

5.2.1 Concepts of the Wage

In the exposition of the model in this section, I will refer to the variable w as the wage, in the sense of the common value of the marginal value of time and the marginal product of labor that would occur in equilibrium in an economy with a market wage with the property that hours of work are chosen on the supply side to equate the marginal value of time to w, and hours of work are chosen on the demand side to equate the marginal product of labor to the same w. This terminology seems most natural for describing the model. On the other hand, I need to distinguish between the supply side and the demand side in the empirical section, because the model includes a wedge separating the marginal value of time and the marginal product of labor. At that point, I will switch to calling the marginal value of time v and the marginal product of labor m.

A related point is that none of these three concepts of the wage is the amount workers receive per hour—they are all shadow concepts reflecting marginal rather than average wages. I use the term *compensation* for actual cash payments to workers. I do not consider data on compensation.

5.2.2 The Family's Decisions

As in most research on choices over time, I assume that preferences are time-separable, though I am mindful of Browning et al.'s (1985) admonition that "the fact that additivity is an almost universal assumption in work on intertemporal choice does not suggest that it is innocuous." In particular, additivity fails in the case of habit.

The family orders levels of hours of employed members, h_t, consumption of employed members, $c_{e,t}$, and consumption of unemployed members, $c_{u,t}$, within a period by the utility function,

$$n_t U(c_{e,t}, h_t) + (1 - n_t) U(c_{u,t}, 0). \qquad (5.1)$$

The family orders future uncertain paths by expected utility with discount factor δ. I view the family utility function as a reduced form for a more complicated model of family activities that includes home production.

The family solves the dynamic program,

$$
\begin{aligned}
V(W_t, \eta_t) \\
= \max_{h_t, c_{e,t}, c_{u,t}} \{ & n_t U(c_{e,t}, h_t) + (1 - n_t) U(c_{u,t}, 0) \\
& + \mathbb{E}_{\eta_{t+1}} \delta V(1 + r_t) \\
& \times [W_t - d(\eta_t) - n_t c_{e,t} - (1 - n_t) c_{u,t}] \\
& - \phi(n_t)(1 - n_t) y_t + w_t n_t h_t, \eta_{t+1}) \}.
\end{aligned}
$$

(5.2)

Here $V(W_t, \eta_t)$ is the family's expected utility as of the beginning of period t, W_t is wealth, and $d(\eta_t)$ is a deduction from wealth depending on the random driving forces η_t that could, for example, arise from the lump-sum component of taxation. The expectation is over the conditional distribution of η_{t+1}. The amount $\phi(n_t)(1 - n_t)$ is the flow of new hires of family members, each of which costs the family y_t.

5.2.3   State Variables

I let λ_t be the marginal utility of wealth (and also marginal utility of consumption):

$$
\lambda_t = \frac{\partial V}{\partial W_t} = \delta(1 + r_t) \mathbb{E}_t \frac{\partial V}{\partial W_{t+1}}.
$$
(5.3)

I take λ_t and the hourly wage w_t as the state variables of the economy relevant to labor-market equilibrium. Both state variables are complicated functions of the underlying exogenous driving forces, η. Marginal utility, λ_t, is an endogenous variable that embodies the entire forward-looking optimization of the household based on its perceptions of future earnings and deductions. The common value of the marginal product of labor and the marginal value of time, w, is an endogenous variable that depends on the amounts of capital and labor used in production, which depend in turn on all of the elements of the labor-market model and on features of the economy not included in that model.

The strategy pursued in the rest of the chapter exploits the property that a vector of four key observable endogenous variables—consumption, hours of work, the employment rate, and the marginal product of labor—are all functions of the two endogenous state variables, λ and w. The four observable variables have a factor structure, with just two latent factors.

5.2.4 Hours, Consumption, and Employment

The family's first-order conditions for hours and the consumption levels of employed and unemployed members are

$$U_h(c_{e,t}, h_t) = -\lambda_t w_t, \tag{5.4}$$

$$U_c(c_{e,t}, h_t) = \lambda_t, \tag{5.5}$$

$$U_c(c_{u,t}, 0) = \lambda_t. \tag{5.6}$$

These conditions define three Frisch functions,

$$c_e(\lambda_t, \lambda_t w_t), \quad h(\lambda_t, \lambda_t w_t), \quad \text{and} \quad c_u(\lambda_t)$$

giving the consumption and hours of the employed and the consumption of the unemployed. I write the functions in this form to connect with research on Frisch labor supply and consumer demand equations. With consumption–hours complementarity, the family assigns a lower level of consumption to the unemployed than to the employed: $c_u < c_e$.

5.3 The Employment Function

Hall (2009) develops the concept of the employment function, $n(\lambda, w)$. It maps the same two variables central to the Frisch household system into the employment rate, 1 minus the unemployment rate. The employment function does not describe family choice alone. Rather, it describes the equilibrium in the labor market resulting from the interaction of workers and employers. Part of the apparatus underlying the employment function comes from Mortensen and Pissarides (1994)—a fairly constant flow of workers out of jobs, a matching function delivering a flow of matches of job seekers to employers depending on unemployment and vacancies, and endogenous recruiting effort with a zero-profit equilibrium for employers. In that model, the factor determining the tightness of the labor market is recruiting effort, which in turn depends on the part of the surplus from a job match that accrues to the employer. In the Mortensen–Pissarides model, the fraction of the surplus going to employers is a constant set by a Nash bargain, so, as Shimer (2005) showed, there is little variation in the amount of the surplus and hence little volatility of unemployment. The employment function allows a more general type of bargaining outcome, where a recession is a time when the share of the surplus accruing to employers declines. I show that a reasonably wide class of bargaining outcomes depends on just the variables λ and w.

Hall and Milgrom (2008, table 3) report that the observed elasticity of the unemployment rate with respect to productivity is about 20. This calculation holds the flow value of nonemployment constant, so it corresponds in the framework of this chapter to holding λ constant. The corresponding elasticity of the employment rate with respect to w is 1.2, the value I use.

I have not found any outside benchmark for the elasticity of $n(\lambda, w)$ with respect to λ. The general view of wage bargaining developed earlier in the chapter does not speak to the value of the elasticity. Accordingly, I choose a value, $\beta_{n,\lambda} = 0.6$, that yields approximately the best fit.

The parameters of the employment function are the only ones chosen on the basis of fit to the aggregate data. Research on search and matching models with realistic nonlinear preferences, non-Nash wage bargaining, and other relevant features has flourished recently and may provide more guidance in the future.

5.4 Econometric Model

I approximate the consumption demands, hours supply, and employment functions as log-linear, with $\beta_{c,c}$ denoting the elasticity of consumption with respect to its own price (the elasticity corresponding to the partial derivative c_1 in the earlier discussion), $\beta_{c,h}$ the cross-elasticity of consumption demand and hours supply, and $\beta_{h,h}$ the own-elasticity of hours supply. I further let $\beta_{n,\lambda}$ denote the elasticity of employment with respect to marginal utility λ and $\beta_{n,w}$ the elasticity with respect to the marginal product w. The model is as follows.

Consumption of the employed:

$$\log c_e = \beta_{c,c} \log \lambda + \beta_{c,h}(\log \lambda + \log w). \qquad (5.7)$$

Consumption of the unemployed:

$$\log c_u = \beta_{c,c} \log \lambda. \qquad (5.8)$$

Hours:

$$\log h = -\beta_{c,h} \log \lambda + \beta_{h,h} (\log \lambda + \log w). \qquad (5.9)$$

Employment rate:

$$\log n = \beta_{n,\lambda} \log \lambda + \beta_{n,w} \log w. \qquad (5.10)$$

Productivity:

$$\log m = \log w + \log \alpha. \qquad (5.11)$$

Table 5.1 summarizes the parameter values I use as the base case. The upper panel gives the elasticities described in the previous section and the lower panel restates them as the coefficients governing the relation (in logs) between the observed variables and the underlying factors, λ and w. The lower panel takes into account the double appearance of λ in the Frisch hours supply and consumption demand functions.

Notice that the employment equation resembles the hours equation, but with larger coefficients. The elasticities of annual hours, nh, with respect to λ and w, are the sums of the coefficients in the second and third lines of the lower panel of table 5.1. The effect of including a substantially elastic employment function is to make annual hours far more elastic than is labor supply in household studies. The introduction of an employment function is a way to rationalize the fact of elastic annual hours with the microeconomic finding that the weekly hours of individual workers are not nearly so elastic. The employment function is not a feature of individual choice, but of the interaction of all workers and all employers.

5.4.1 Long-Run Properties and Detrending

Hours of work, h, have been roughly constant over the past sixty years. Given constant hours, the family's

Table 5.1. Parameters and corresponding coefficients in the equations of the model.

Elasticities

Consumption with respect to λ	-0.5
Consumption with respect to λw	0.3
Hours with respect to λ	-0.3
Hours with respect to λw	0.7
Employment with respect to λ	0.6
Employment with respect to w	1.2

Coefficients

	λ	w
Consumption	-0.2	0.3
Hours	0.4	0.7
Employment	0.6	1.2
Productivity	0	1

budget constraint requires, roughly, that consumption grow at the same rate as the marginal product of labor, w. Putting these conditions into the equations above yields the standard conclusion that the own-price elasticity of consumption demand, $\beta_{c,c}$, is -1 (log preferences) and that the cross-effect, $\beta_{c,h}$, is 0. Neither of these conditions is consistent with evidence from household studies. Therefore, I interpret the model as describing responses at cyclical frequencies but not at low frequencies, where trends in household technology and preferences come into play. Thus I study detrended data, specifically, the residuals from regressions of the data, in log form, on a third-order polynomial in time. Figure 5.1 showed the detrended data.

The detrending also removes the production elasticity, α, which I assume moves only at low frequencies.

The uncompensated hours supply function is backward-bending for the parameter values I use. By uncompensated, I mean subject to a budget constraint where consumption equals the amount of earnings, wh. Solving equations (5.7) and (5.9) for the change in h and c for a doubling of w subject to constancy of $\log(wh/c)$, I find that hours would fall by 15% and consumption would rise by 70%. With preferences satisfying the restriction of zero uncompensated wage elasticity, hours would remain the same and consumption would double.

My approach here is the opposite of Shimer's (2008). He requires that preferences satisfy the long-run restrictions and therefore does not match the elasticities I use. Because preferences are a reduced form for a more elaborate specification including home production, where productivity trends might logically be included, it is a matter of judgment whether to impose the long-run restrictions. Of course, the best solution would be a full treatment of the household with explicit technology and measured productivity trends.

5.4.2 Consumption

The model disaggregates the population by the employed and unemployed, who consume c_e and c_u respectively. Only average consumption c is observed. It is the average of the two levels, weighted by the employment and unemployment fractions:

$$c = nc_e + (1 - n)c_u. \qquad (5.12)$$

I solve this equation for c_e given the hypothesis that the consumption of the unemployed is a fraction ρ of the consumption of the employed:

$$c_e = \frac{c}{n + (1 - n)\rho}. \tag{5.13}$$

The effect of this calculation is to remove from c the mix effect that occurs when employment falls and more people are consuming the lower amount c_u. The adjustment is quite small. I drop c_u from the model because it is taken to be strictly proportional to c_e.

The evidence discussed in chapter 2 suggests that $\rho = 0.85$.

5.4.3 Disturbances and Their Variances

The data do not fit the model exactly. I hypothesize additive disturbances ϵ_c, ϵ_h, ϵ_n, and ϵ_m in the equations for the four observed variables. I assume that these are uncorrelated with λ and w. This assumption is easiest to rationalize if the ϵs are measurement errors.

See Hall (2009) for a discussion of the estimation of the variances of the disturbances associated with the four observed variables and the variances and covariance of λ and w, together with the inference of the time series for the disturbances, λ and w.

5.4.4 Restatement of the Model

The model as estimated is as follows.

Consumption of the employed:

$$\log c_e = \beta_{c,c} \log \lambda + \beta_{c,h}(\log \lambda + \log v) + \log \epsilon_c. \tag{5.14}$$

Hours:

$$\log h = -\beta_{c,h} \log \lambda + \beta_{h,h}(\log \lambda + \log v) + \log \epsilon_h. \tag{5.15}$$

Employment rate:

$$\log n = \beta_{n,\lambda} \log \lambda + \beta_{n,v} \log v + \log \epsilon_n. \tag{5.16}$$

Table 5.2. Covariances, standard deviations, and correlations of logs of consumption, hours, employment, and productivity.

	(i)	(ii)	(iii)	(iv)
Covariances ×10,000				
Consumption of employed	1.27	−0.12	0.36	1.46
Hours per worker		1.11	0.96	−0.03
Employment rate			1.61	0.48
Productivity				2.80
Standard deviations (%)	1.13	1.05	1.27	1.67
Correlations				
Consumption of employed	1.00	−0.10	0.25	0.78
Hours per worker		1.00	0.72	−0.01
Employment rate			1.00	0.23
Productivity				1.00

(i) Consumption of employed; (ii) hours per worker; (iii) employment rate; (iv) productivity.

Productivity:
$$\log m = \log v + \epsilon_m. \tag{5.17}$$

5.5   Properties of the Data

For information about the data, see Hall (2009) and the further information available in connection with that paper on my website.

Table 5.2 shows the covariance and correlation matrixes of the logs of the four detrended series. Consumption is correlated positively with employment—it is quite procyclical. Consumption–hours complementarity can explain this fact. Not surprisingly, hours and employment are quite positively correlated.

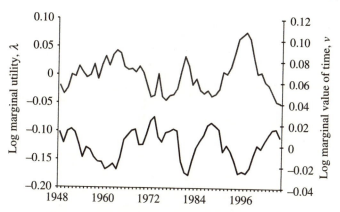

Figure 5.2. Inferred values of marginal utility and marginal value of time.

Consumption also has by far the highest correlation with productivity.

The standard deviation of the employment rate is about 25% higher than the standard deviation of hours—the more important source for the added total hours of work in an expansion is the reduction in unemployment. Hours are not very correlated with productivity. Note that productivity has the highest standard deviation of the four variables—amplification of productivity fluctuations need not be part of a model in which productivity is the driving force.

5.6 Results

Figure 5.2 shows the estimates of detrended log marginal utility, $\log \lambda$, and marginal value of time, $\log v$. The figure shows the pronounced negative correlation (-0.81) between marginal utility and the marginal value of time.

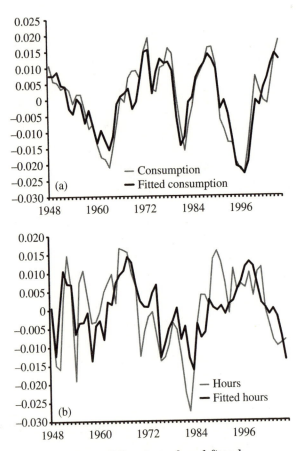

Figure 5.3. Actual and fitted
values of the four observables.

5.6.1 Fitted Values for Observables

Figure 5.3 shows the fitted values for the four observables from the time series for λ and v, using the coefficients in the bottom panel of table 5.1. The two-
factor setup is highly successful in accounting for
the observed movements of all four variables. Little

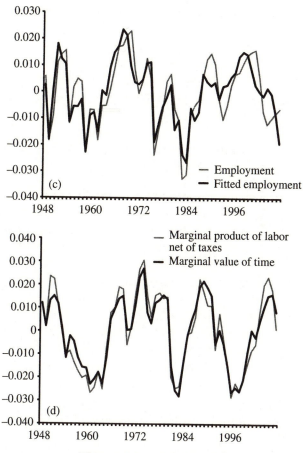

Figure 5.3. *Continued.*

is left to the idiosyncratic disturbances. Of course, two factors are likely to be able to account for most of the movement of four macro time series, especially when two pairs of them, hours-employment and consumption-productivity, are fairly highly correlated. But the choices of the factors and the factor loadings are not made, as in principal components, to provide the best match. The loadings are based in part on preference parameters drawn from earlier research. The success of the model is not so much the good fit shown in the figure, but rather achieving the good fit with coefficients that satisfy economic reasonability.

5.6.2 Reconciliation of the Marginal Value of Time and the Marginal Product of Labor

Figure 5.3(d) shows the extent that the model is able to generate estimates of the marginal value of time that track data on the tax-adjusted marginal product of labor. The marginal value of time follows measured productivity quite closely. The figure does not support any diagnosis of repeated or severe private bilateral inefficiency. Of course, one of the major factors accounting for the absence of bilateral inefficiency is to shift the efficiency issue from the bilateral situation of a worker and an employer to the economy-wide situation of job seekers and employers interacting collectively and anonymously. Though the model portrays the movements in figure 5.3 as privately bilaterally efficient, it does not portray socially efficient allocations.

The biggest departure from the normal view of the U.S. business cycle in the value of time–productivity plot occurred in the middle of the 1990s, usually viewed as a time of full employment and normal conditions, but portrayed here as an extended period of low productivity matched to low marginal value of time. Low productivity, trend- and tax-adjusted, comes straight from the data. How does the model infer that the marginal value of time was equally depressed?

Figure 5.3(a) shows that consumption was low, so marginal utility was high. Both hours and employment respond positively to λ—people work harder when they feel poorer, according to the standard theory of the household—and the employment function also responds positively—the labor market is tighter when λ is higher. Thus the slump in the mid 90s was a time when people worked hard because they did not feel well off. It was not a recession in the sense of a period of a slack labor market but it was a transitory period of depressed productivity and depressed value of time.

5.7 Concluding Remarks

A model that combines a standard Frisch consumption demand and hours supply system with a generalized Mortensen-Pissarides employment function gives a reasonably complete accounting for the cyclical movements of four key variables: consumption, weekly hours, the employment rate, and productivity. The employment rate, not a feature of family decision making alone, accounts for well over half of the observed movements in labor input. Family decisions matter directly for consumption and weekly hours and enter the determination of the employment rate through the household's role in the wage bargain.

6

Idiosyncratic Risk

Like chapter 3, this chapter considers the burden on the individual or family from lack of insurance. In chapter 3, people did not buy insurance even though it was available—and even subsidized, in the case of women. In this chapter, insurance is nonexistent, even though the risk is enormous and the payoff to insurance would be gigantic if a market were possible. For good reasons, no insurance is possible.

The chapter is about the risk that an entrepreneur in a high-tech startup faces. An entrepreneur's primary incentive is ownership of a substantial share of the enterprise that commercializes the entrepreneur's ideas. An inescapable consequence of this incentive is the entrepreneur's exposure to the idiosyncratic risk of the enterprise. Diversification or insurance to ameliorate the risk would necessarily weaken the incentives for success.

The startup companies studied here are backed by venture capital. These startups are mainly in information technology and biotechnology. They harness teams comprising entrepreneurs (scientists, engineers, and executives), venture capitalists (general partners of venture funds), and the suppliers of capital (the limited partners of venture funds). During the startup process, entrepreneurs collect only submarket salaries. The compensation that attracts them to start-ups is the share they receive of the value of a company if it goes public or is acquired.

For much more information about the venture-capital process and the data, see Hall and Woodward (2007, 2009).

The most important finding is that the reward to the entrepreneurs who provide the ideas and long hours of hard work in these startups is zero in almost three quarters of the outcomes, and small on average once idiosyncratic risk is taken into consideration.

Although the average ultimate cash reward to an entrepreneur in a company that succeeds in landing venture funding is $3.6 million, most of this expected value comes from the small probability of a great success. An individual with a coefficient of relative risk aversion of 2 and assets of $140,600 is indifferent between employment at a market salary and entrepreneurship. With lower risk aversion or higher initial assets, the entrepreneurial opportunity is worth more than alternative employment. Entrepreneurs are drawn differentially from individuals with lower risk aversion and higher assets. Other types of people that may be attracted to entrepreneurship are those with preferences for that role over employment and those who exaggerate the likely payoffs of their own products. The model does not include these factors, however—it uses standard preferences based on consumption levels alone.

The analysis focuses on the joint distribution of the duration of the entrepreneur's involvement in a start-up—what we call the venture lifetime—and the value that the entrepreneur receives when the company exits the venture portfolio. Exits take three forms: (1) an

initial public offering, in which the entrepreneur receives liquid publicly traded shares or cash (if she sells
her own shares at the IPO or soon after) and has the
opportunity to diversify; (2) the sale of the company to
an acquirer, in which the entrepreneur receives cash or
publicly traded shares in the acquiring company and
has the opportunity to diversify; and (3) shutdown or
other determination that the entrepreneur's equity interest has essentially no value. Most IPOs return substantial value to an entrepreneur. Some acquisitions
also return substantial value, while others may deliver
a meager or zero value to the entrepreneur.

The joint distribution shows a distinct negative
correlation between exit value and venture lifetime.
Highly successful products tend to result in IPOs or acquisitions at high values relatively quickly. These outcomes are favorable for entrepreneurs in two ways.
First, the value arrives quickly and is subject to less
discounting. Second, the entrepreneur spends less
time being paid a low startup salary and correspondingly more time with higher post-startup compensation, in the public version of the original company, in
the acquiring company, or in another job. A fraction of
entrepreneurs launch new startups after exiting from
an earlier startup.

The chapter studies exit values from the point of
view of the individual entrepreneur. About a quarter of entrepreneurs do not share the proceeds with
other entrepreneurs; they operate solo. Another quarter share the entrepreneurial role equally with another
founder. In the remaining cases, entrepreneurial ownership is distributed asymmetrically between a pair of
entrepreneurs or there are three or even more entrepreneurs. The total entrepreneurial return delivered
by each venture company allows an inference of the

returns to the individual entrepreneurs from information about the distribution among entrepreneurs. All of the tabulations in the chapter refer to entrepreneurs, not to companies.

A family dynamic program delivers a unified analysis of the factors affecting the entrepreneur's risk-adjusted payoff. The analysis takes account of the joint distribution of exit value and venture lifetime and of salary and compensation income. It allows the calculation of the certainty-equivalent value of the entrepreneurial opportunity—the amount that a prospective entrepreneur would be willing to pay to become a founder of a venture-backed startup. For a risk-neutral individual, the certainty-equivalent is $3.6 million. With mild risk aversion and savings of $100,000, however, the amount is only $0.7 million and with normal risk aversion and that amount of savings, the certainty-equivalent is slightly negative.

6.1 The Joint Distribution of Lifetime and Exit Value

The lifetime of a startup—the time from inception to the entrepreneurs' receipt of cash from an exit event—plays a key role in the analysis. Entrepreneurs prefer short lifetimes for two reasons. First, their salaries at a venture-backed startup are modest; they forgo a full return to their human capital during the lifetime. Second, the time value of money places a higher value on cash received sooner.

Lifetimes and exit values are not distributed independently. In particular, a substantial fraction of startups linger for many years and then never deliver much cash to their founders. And some of the highest exit values occurred for companies like YouTube that exited soon after inception.

The calculations also need to make the transition from data based on companies to distributions over entrepreneurs and to account for companies that have not yet exited. Hall and Woodward (2009) describe how this is done.

We take a flexible view of the joint distribution, as appropriate for our rich body of data. We place lifetimes τ and values v in 9 and 11 bins respectively and estimate the 99 values of the joint distribution defined over the bins. Table 6.1 shows the joint distribution in these bins.

6.2   Economic Payoffs to Entrepreneurs

Venture-backed companies typically have a scientist or similar expert, or a small group, who supply the original concept, contribute a small amount of capital, and find investors to supply the bulk of the capital. These entrepreneurs, together with any angels, own all of the shares in the company prior to the first round of venture funding.

The entrepreneurs are specialized in ownership of the venture-stage firm. The approach to valuation takes account of the heavy exposure of the entrepreneur to the idiosyncratic volatility of the company. It also takes account of the modest salaries that entrepreneurs generally receive during the venture phase of the development of their companies and of the lifetime of the company, which affects the discounting applied to the exit value and the burden of the low salary.

Table 6.1. Joint distribution of venture lifetime and exit value, percent probability by cell.

Exit value (millions of dollars)	Venture lifetime									All
	0 to 1	1 to 2	2 to 3	3 to 4	4 to 5	5 to 6	6 to 7	7 to 9	10+	
0	6.14	12.49	11.50	9.78	8.37	6.22	4.91	6.48	7.23	73.1
0 to 3	0.224	0.729	0.757	0.931	0.861	0.736	0.624	0.877	0.929	6.67
3 to 6	0.187	0.435	0.459	0.521	0.519	0.401	0.371	0.410	0.413	3.72
6 to 12	0.243	0.569	0.619	0.644	0.587	0.454	0.390	0.438	0.416	4.36
12 to 20	0.219	0.471	0.498	0.489	0.426	0.321	0.250	0.297	0.284	3.26
20 to 50	0.425	0.739	0.800	0.714	0.599	0.442	0.349	0.397	0.323	4.79
50 to 100	0.238	0.481	0.435	0.338	0.249	0.149	0.132	0.151	0.106	2.28
100 to 200	0.146	0.297	0.269	0.183	0.099	0.052	0.028	0.061	0.042	1.18
200 to 500	0.070	0.140	0.121	0.090	0.020	0.016	0.006	0.024	0.023	0.511
500 to 1,000	0.0145	0.0392	0.0124	0.0103	0.00176	0.00581	0.00194	0.00204	0.00000	0.088
1,000+	0.00464	0.00990	0.01103	0.00389	0.00000	0.00603	0.00000	0.00000	0.00000	0.035
All	7.91	16.41	15.48	13.70	11.73	8.80	7.07	9.14	9.77	100

The model assumes that the entrepreneurs in a company have already made all of their financial investments in their company; all further funds will come from venture investors. This assumption seems to be generally realistic, though of course some entrepreneurs are able to continue financing their companies alongside venture investors. An entrepreneur has some savings available to finance consumption beyond what the relatively low venture salary will support. Entrepreneurs cannot borrow against future earnings or against the possible exit value of the company. This assumption is entirely realistic. Thus the entrepreneur makes a decision each year about how much to draw down savings during the year; that is, by how much consumption will exceed the venture salary.

6.2.1 Analytical Framework

The framework starts from a standard specification of intertemporal preferences for entrepreneurs—they order random consumption paths according to

$$\mathbb{E} \sum_t \left(\frac{1}{1+r} \right)^t u(c_t). \tag{6.1}$$

Here r is the entrepreneur's rate of time preference and the rate of return on assets; $u(c)$ is a concave period utility. We define the function $U(W)$ as the utility from a constant path of consumption funded by wealth W:

$$U(W) = \frac{1+r}{r} u\left(\frac{r}{1+r} W \right). \tag{6.2}$$

The multiplication by

$$\frac{1+r}{r}$$

turns flow utility into discounted lifetime utility. The quantity

$$\frac{r}{1+r} W$$

is the flow of consumption to be financed by the return on the wealth at rate r.

The variable *wealth*, W_t, measures the entrepreneur's total command over resources, and so incorporates the expected value of future compensation (human wealth), while *assets*, A_t, denotes holdings of nonhuman wealth as savings. A_t does not include the entrepreneur's holdings of shares in the startup. For an entrepreneur in year t of a startup that has not yet exited, $W_t(A_t)$ is the *wealth-equivalent* of the entrepreneur's command over resources, counting what remains of the entrepreneur's original nonhuman wealth, A_t, and the entrepreneur's random future payoff from the startup, conditional on not having exited to this time. The definition is implicit: $U(W_t(A_t))$ is the expected utility from maximizing (6.1) over consumption strategies.

Now let $U(W_t(A_t))$ be the value, in utility units, associated with an entrepreneur in a nonexited company t years past venture funding, as a function of current nonentrepreneurial assets A_t. The model could use a value function $U_t(A_t)$ without interposing the function $W_t(A_t)$. With $W_t(A_t)$ as the value function, the model needs to incorporate the concave transformation $U(W_t(A_t))$ so that the Bellman equation adds up utility, according to the principle of expected utility. The slightly roundabout approach of stating the findings in terms of the wealth-equivalent $W_t(A_t))$ makes the units meaningful, whereas the units of utility are not. Furthermore, in the benchmark case, utility is negative, a further source of confusion. Note that W captures initial assets, venture salary, venture exit value, and subsequent compensation in a post-venture position, when it is calculated at time zero for an entrepreneur.

The company has a conditional probability or hazard π_t of exiting at age t. At exit, it pays a random amount X_t to the entrepreneur. Upon exiting, the entrepreneur's value function is $U(W^*(A))$, where A now includes the cash exit value. The entrepreneur's consumption is limited by assets left from the previous year—no borrowing against future earnings may occur. The entrepreneur's dynamic program is

$$
\begin{aligned}
U(W_t) \\
= \max_{c_t < A_t} \Bigg[& u(c_t) \\
& + \frac{1}{1+r}(1 - \pi_{t+1}) \\
& \qquad \times U(W_{t+1}((A_t - c_t)(1+r) + w)) \\
& + \frac{1}{1+r}\pi_{t+1}\,\mathbb{E}_X \\
& \qquad \times U(W^*((A_t - c_t)(1+r) + X_{t+1})) \Bigg].
\end{aligned}
$$
(6.3)

The post-venture value function is

$$
U(W^*(A)) = \frac{1+r}{r}\,u\!\left(\frac{rA + w^*}{1+r}\right). \tag{6.4}
$$

Here w^* is post-venture compensation, including employee stock options, at the nonventure continuation of this company or another company. From equations (6.2) and (6.4), we have

$$
W^*(A) = A + \frac{w^*}{r}. \tag{6.5}
$$

Note that this is additive in A. But when future earnings are random, the entrepreneur's risk aversion enters the calculation of the wealth-equivalent.

Each of the value functions $U(W_t(A_t))$ is piecewise linear with 500 knots between zero and \$49 million, spaced exponentially. Calculation is by backward recursion (value function iteration). The utility function has constant relative risk aversion, y. The base case is $y = 2$, a venture salary w equal to the post-tax value of \$150,000, post-venture compensation w^* equal to the post-tax value of \$300,000, and starting assets of $A_0 = \$1$ million.

A useful feature of the wealth-equivalent is that the difference between its value for an entrepreneur with given initial assets and its value for an individual who holds a nonventure position paying w^* and with the same initial assets is the amount that the second would be willing to pay to become an entrepreneur, the *certainty-equivalent value of the entrepreneurial opportunity*, denoted $\tilde{A}$. This property follows from the additivity of the nonentrepreneurial wealth-equivalent we noted earlier. Chapter 4 noted that Brown and Finkelstein could have set their insurance problem up this way.

6.2.2 Results

Figure 6.1 shows $W_0(A_0)$, the wealth-equivalent for an entrepreneurial experience as of its beginning and $W^*(A_0)$, the wealth-equivalent for a nonentrepreneur, both as functions of the common value of their initial assets, shown on the horizontal axis. The certainty-equivalent value of the venture opportunity is the vertical difference between the two curves. The non-entrepreneurial value is a straight line with unit

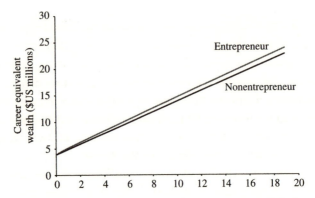

Figure 6.1. Certainty-equivalent career wealth for
entrepreneurs and nonentrepreneurs.

slope—a dollar of extra initial assets becomes a dollar of wealth, because we assume that the nonventure
individual faces no uncertainty. On the other hand, a
dollar of extra initial assets becomes more than a dollar of equivalent wealth, because initial wealth has no
uncertainty and thus dilutes the uncertainty from the
venture outcome. This property is a cousin of the principle that people should treat risky outcomes as if they
were worth essentially their expected values, when the
outcomes are tiny in relation to their wealth. The slope
of the entrepreneur's value is more than 3 at low levels
of assets but declines to 1.03 at assets of $20 million.

The figure shows that, despite the chance of making
hundreds of millions of dollars in a startup, the economic advantage of entrepreneurship over an alternative career is not substantial. The burden of the idiosyncratic risk of a startup falls most heavily on those
with low initial assets. The entrepreneur with less than
a million dollars of initial assets faces a heavy burden
from the risk and has a lower career wealth than the
nonentrepreneur.

Table 6.2. Certainty-equivalent
value of the venture opportunity.

| | | CEEO (millions of dollars) | | | |
| | | Assets at beginning (millions of dollars) | | | |
y	ξ	0.1	1	5	20
0	300	3.6	3.6	3.6	3.6
0	600	2.9	2.9	2.9	2.9
0	2,000	0.1	0.1	0.1	0.1
0.9	300	0.7	0.9	1.2	1.6
0.9	600	−0.2	0.3	0.7	1.0
0.9	2,000	−5.8	−4.0	−2.2	−1.7
2	300	−0.1	0.3	0.6	1.1
2	600	−1.8	−0.5	0.0	0.6
2	2,000	−13.8	−8.8	−3.6	−2.1

y is the coefficient of relative risk aversion; ξ is the pretax compensation at the nonentrepreneurial job (in thousands of dollars per year); CEEO, certainty-equivalent of entrepreneurial opportunity.

Table 6.2 gives the certainty-equivalent value of the entrepreneurial opportunity for thirty-six combinations of the three determinants: the coefficient of relative risk aversion, the compensation at an alternative, nonentrepreneurial job, and the entrepreneur's assets at the beginning of entrepreneurship. The first three lines take the entrepreneur to be risk neutral, so the values are just present values at the 5% annual real discount rate. In this case, the value is the same for any level of initial assets. The value is $3.6 million. The value is $2.9 million for an individual with a nonentrepreneurial opportunity to earn $600,000 per year before tax. If the nonentrepreneurial opportunity pays $2 million per year before tax, the venture opportunity has barely positive value. A typical startup probably cannot attract an established top executive from a large public corporation, even if the executive is risk neutral, as their earnings are generally even higher than $2 million.

The conclusions from the table are similar if the individual is mildly risk averse, with a coefficient of relative risk aversion of 0.9. The advantage of the entrepreneurial opportunity, stated as a wealth-equivalent, is only $0.7 million for an entrepreneur with $0.1 million in assets and only $1.2 million for an entrepreneur with $5 million. These figures are negative or only slightly positive if the nonentrepreneurial opportunity pays $600,000 per year before tax.

At the standard value of 2 of the coefficient of relative risk aversion from chapter 2, the advantage of the entrepreneurial opportunity is generally small or negative—deeply negative if the nonentrepreneurial opportunity pays $2 million per year. In the base case, with nonentrepreneurial compensation of $300,000 per year before tax and $1 million in assets, the advantage of the entrepreneurial opportunity is only $0.3 million. The incentive is not impressive for larger asset holdings. With higher compensation at the non-entrepreneurial job, the advantage disappears unless the individual is quite rich.

6.3 Entrepreneurs in Aging Companies

The discussion so far has focused on the risk-adjusted payoff to a potential entrepreneur at the decision

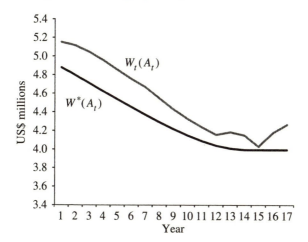

Figure 6.2. Entrepreneurial value and nonentrepreneurial value prior to exit.

point when venture funding first becomes available. This section considers the same issue at later decision points, as the startup ages. The discussion is conditional on the company not having exited.

The dynamic program of equation (6.3) assigns a value $W_t(A_t)$ to the entrepreneur's position in each year t that the company has not exited. The path is the same for all companies, under the assumptions of the model. The entrepreneur's value falls as the company ages for two reasons. First, the entrepreneur generally consumes out of assets, so assets decline. Second, early exits are the most valuable exits, so aging another year means that the remaining potential exit values are not as valuable. Figure 6.2 shows the path of $W_t(A_t)$. It declines from $5.2 million at the outset to $4.3 million at age 10, conditional on no exit. From that point the value rises, because the distribution of

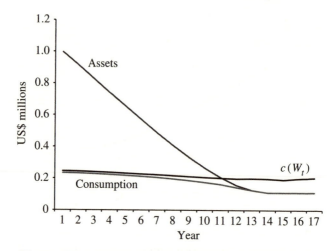

Figure 6.3. Consumption and assets prior to exit.

exit values becomes more favorable, though not as favorable as for young startups.

The figure also shows the individual's value of a nonentrepreneurial job, $W^*(A_t)$. It declines as well, but only for the first reason, the drawdown of assets to finance consumption in excess of the low startup salary.

Figure 6.3 shows the paths of assets and consumption as a company ages. For the first decade, assets decline because consumption exceeds the modest startup salary and the entrepreneur has no other source of current cash, pending a favorable exit. During this period consumption declines, because, as an exit fails to occur during the early years, the entrepreneur learns that risk-adjusted well-being, as measured by $W_t(A_t)$, has declined. Eventually, assets fall to the point of consumption. From this point until exit, the entrepreneur lives on the salary and maintains assets only as a way to spread consumption between paychecks (we assume, for simplicity, that the entrepreneur receives the salary at the end of each year and we measure assets at the beginning of the year). The line labeled $c(W_t)$ shows the level of consumption that a consumer without a cash-flow constraint would choose, given lifetime prospects as measured by W_t. Consumption starts out only slightly below this level, but as the entrepreneur depletes assets, consumption falls toward the cash-flow limit. In the event that the startup ages into its second decade, the cash-flow constraint keeps consumption far below its unconstrained level.

6.4 Concluding Remarks

The contract between venture capital and entrepreneurs does essentially nothing to alleviate their financial extreme specialization in their own companies. Given the nature of the gamble, entrepreneurs would benefit by selling some of the value that they would receive in the best outcome on the right, when they would be seriously rich, in exchange for more wealth in the most likely of zero exit value on the left. It would be hard to find a more serious violation of the Borch–Arrow optimality condition—equality of marginal utility in all states of the world—than in the case of entrepreneurs.

A diversified investor would be happy to trade this off at a reasonable price, given that most of the risk is idiosyncratic and diversifiable. But venture capitalists will not do this—they do not buy out startups at the early stages and they do not let entrepreneurs pay themselves generous salaries. They use the exit value as an incentive for the entrepreneurs to perform their jobs. Moral hazard and adverse selection bar the provision of any type of insurance to entrepreneurs—they must bear the huge risk.

The venture capital institutions of the United States convert ideas into functioning businesses. We show that the process contains an important bottleneck—for good reasons based mainly on moral hazard, the venture contract cannot insure entrepreneurs against the huge idiosyncratic risk of a startup. Risk-adjusted payoffs to the entrepreneurs of startups are remarkably small. Although our results are based entirely on the venture process, we believe that no other arrangement is much better at solving the problem of getting smart people to commercialize their good ideas.

7

Financial Stability with Government-Guaranteed Debt

7.1 Introduction

In modern economies, the government guarantees the debt of many borrowers. In a few cases, the promise is explicit; in others it is implicit but known to be likely; and in others, the guarantee occurs because the alternative is immediate collapse, with substantial harm to the rest of the economy. The modern government cannot stop itself from making good on the obligations of many borrowers, large and small. I demonstrate that debt guarantees deplete equity from firms at times of declines in asset values. Not only do firms fail to replace equity lost when leveraged portfolios lose value, but they have an incentive to deplete equity further, by paying unusually high dividends.

The government adopts a safeguard to protect the taxpayers against the worst abuses of guarantees—it imposes a capital requirement to limit the ratio of guaranteed debt to the value of the underlying collateral.

In the United States, organizations with explicit guarantees on some debt (deposits) are mainly banks. The nondeposit obligations of banks and other intermediaries, notably the two huge mortgage-holders Fannie Mae and Freddie Mac, enjoy market values that only make sense on the expectation of a government guarantee. The recent interventions to avert the collapse of Bear Stearns and AIG confirmed that the government will pay off on private debt obligations in times of stress even in sectors distant from any formal debt guarantees. The Federal Housing Administration guarantees the debt of individual mortgage borrowers; the government is extending this guarantee to a larger set of individual borrowers under pressure from declining housing prices. Almost any borrower faces some probability that adverse future events will result in the government repaying the borrower's debt.

A key issue is the withdrawal of equity capital from firms with guaranteed debt—the phenomenon I call *equity depletion*. A counterpart is the unwillingness of investors to supply equity to firms that face positive probabilities of insolvency and payoffs on government guarantees. Equity depletion rises along with the probability of default. Withdrawing equity from a firm in one period has zero marginal cost in states next period where default occurs and the government pays off on guaranteed debt—the value of equity claims is zero in those states. If a larger fraction of future states has zero equity value, the expected payoff to equity investments this period declines. In a partial equilibrium setting, this factor would result in knife-edge behavior—the owners of a firm would pay out all of the equity value of a firm as a dividend so as to maximize the value of the government bailout. Akerlof and Romer (1993) describe actions of this type leading up to the savings and loan crisis in the United States in the 1980s. In the model of this chapter, however, a countervailing force limits the depletion of capital.

Any injection of equity to firms in general comes from reduced consumption and any removal of capital takes the form of a consumption binge. With a nonzero value of the elasticity of marginal utility with respect to consumption, the desire to smooth consumption keeps equity flows into and out of firms at finite rates. But consumption does rise substantially in periods when expected defaults and accompanying bailouts become more likely.

The chapter reaches these conclusions in a simple general-equilibrium model. Capital is the only factor of production; output is proportional to capital. The real return to capital is a random variable. Each period, consumers decide between consuming and saving. The government guarantees debt secured by capital. If the borrower's capital falls far enough, because of negative returns, the firm may default because its quantity of capital falls below the amount of its debt.

I characterize the limitations on the government's debt guarantee in what I believe is a realistic way. The government enforces a capital requirement. At the time a company issues debt, the amount borrowed may not exceed a specified fraction of the firm's capital. The remaining value is the borrower's equity, sufficient to meet the capital requirement. If the return is negative enough so that the new quantity of capital falls short of the value of the debt, the government makes up the difference. The lender receives a payment of the difference. Equity shareholders in the firm receive nothing back when default occurs and the government pays off.

My characterization of the government's capital requirement has an important dynamic element. If the quantity of capital falls, but not enough to push the borrower into insolvency, the borrower may keep debt at its earlier level. The government fails to follow the principle of prompt corrective action. Under that principle, the borrower would mark its collateral to market and could borrow only the specified fraction of the new, lower value of the collateral. Instead, the government acts as if the collateral had its historical value and permits the borrower to keep the historical level of debt, which the government guarantees.

Figure 7.1 shows the operation of the model in an example of fifty years of experience. Figure 7.1(a) shows the cumulative return to capital, the exogenous driving force of the model, the only source of departures from a smooth growth path. This part also shows a key variable of the model, the fraction of the value of capital financed by equity. In periods when the return is positive, the fraction is about 30%, reflecting the government's capital requirement of that amount. But when returns are negative, equity depletion occurs—the equity fraction declines. One reason is the government's rule that firms may keep debt at its previous level even capital declines. Capital requirements are based on the book value of assets, not the current value. But another reason is the incentive for firms to increase their payouts when default looms.

In the fifty-year history shown in the figure, default occurs twice. In the first one, several consecutive negative returns cause severe equity depletion, followed by default. In the second default, consecutive shocks deplete equity less severely, and equity depletion is smaller, but default does occur. In the middle of the fifty years, a group of negative returns depresses equity to about 5% of capital but does not cause default.

Figure 7.1(b) shows the resulting volatility in the consumption/capital ratio and in the interest rate on

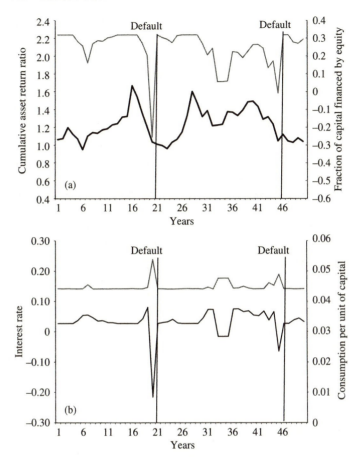

Figure 7.1. Example of a history from the model. (a) Cumulative asset return ratio (black line, left axis), fraction of capital financed by equity (gray line; right axis). (b) Interest rate (black line, left axis), consumption per unit of capital (gray line, right axis).

safe debt. Consumption rises whenever equity depletion occurs from negative returns. Consumption remains high as long as equity is positive and below

normal. Consumers perceive that extracting equity
from firms and consuming it may be free in a year
when default is unusually likely, because the increased
guarantee pays for the consumption should default
occur. Consumption is strongly mean-reverting in this
economy. When consumption is high, it will probably
decline next year either because a positive return returns equity to normal or because a negative return
will cause default, in which case consumption also
returns to normal. Thus periods of depleted equity
are also periods of negative expected consumption
growth. The interest rate tracks expected consumption growth. As the figure shows, the interest rate
falls dramatically to negative levels during periods of
depleted equity.

It should go without saying that the model in this
chapter falls short of capturing reality. It makes no
claim to portray the actual events in financial markets in 2007 and 2008. Rather, it is a full working out
of the implications in a fairly standard model of one
important feature of financial markets, widespread
government guarantees of debt.

7.2 Options

Black and Scholes (1973) identified an incentive for equity depletion for a firm with nonguaranteed debt. In
that situation, the shareholders hold a call option on
the firm's assets with an exercise price equal to the
face value of the debt. If the firm depletes equity by,
for example, paying a dividend of $1, the sharehold-
ers gain $1 in hand but lose less than $1. The δ of
the option—the derivative of its value with respect to
the value of the firm's assets—is less than 1. Hence it

would appear that the shareholders have an unlimited incentive to deplete equity. But every dollar the shareholders gain is a dollar lost by the bondholders. Black and Scholes went on to observe that bondholders have contractual protections against this conduct.

Equity depletion without a debt guarantee is the result of exploitation of bondholders by shareholders. If investors all held shares and bonds in the same proportion, the incentive for equity depletion would not exist. Equity depletion arising from the incentive identified by Black and Scholes is the result of agency frictions between shareholders and bondholders. This chapter is *not* about that incentive. I assume a frictionless solution to the potential agency problem.

With guaranteed debt, the Black–Scholes analysis continues to hold—paying out $1 in dividends lowers the value of the shareholders' call option by less than $1. The guarantee immunizes the bondholders from any loss, however. The shareholders capture extra value not from the bondholders but from the government. Equity depletion appears to be an unlimited opportunity to steal from the government.

Just as the bondholders have at least some contractual protection in the case of nonguaranteed debt, the government, in the model developed here, imposes restrictions on equity depletion. In particular, firms cannot steal directly by paying out such high dividends that debt exceeds the current value of assets. Firms may only gamble that a decline in asset values next period will trigger a guarantee payment. But it turns out that this constraint rarely binds. In the aggregate, equity depletion requires increased consumption. With a reasonable value of the elasticity of intertemporal substitution, consumers' desire to smooth consumption limits equity depletion, though equilibrium consumption in the model is quite volatile.

7.3 Model

7.3.1 Basic Stochastic Growth Model

Capital K is the only factor of production. The return ratio for capital is the random variable A, so one unit of capital becomes A units of output at the beginning of the next period. The return ratio is exogenously, independently, and identically distributed over time. I will refer to a value of A greater than 1 as a positive return and a value below 1 as a negative return. I will also refer to a value below 1 as reducing the value of capital. Each person consumes c. I let a prime denote next period's value of a variable. I denote state variables with a hat. The state variable associated with capital is the quantity of output available for reinvestment or consumption, $\hat{K}$. Its law of motion is

$$\hat{K}' = A(\hat{K} - c). \tag{7.1}$$

Consumers have power utility functions with coefficient of relative risk aversion of y and intertemporal elasticity of substitution of $1/y$. Consumers solve the dynamic program,

$$V(\hat{K}) = \max_c \frac{c^{1-y}}{1-y} + \beta \, \mathbb{E}_A V(\hat{K}'). \tag{7.2}$$

Here β is the consumer's discount ratio. The value function takes the form

$$V(\hat{K}) = V\hat{K}^{1-y}. \tag{7.3}$$

so the dynamic program becomes

$$V\hat{K}^{1-y} = \max_c \frac{c^{1-y}}{1-y} + \beta V \, \mathbb{E}_A \hat{K}'^{1-y}. \tag{7.4}$$

The solution for the constant V does not have a closed form but is unique and computationally benign. Notice that one could divide both sides by $\hat{K}^{1-\gamma}$ to eliminate capital as a state variable, replacing c with $\tilde{c} = c/\hat{K}$. Thus the model implies a constant consumption/capital ratio.

Throughout the chapter, I measure all values in consumption units.

7.3.2 Firms and Consumers

The basic growth model has constant returns to scale, so the boundaries of the firm are indeterminate. Without loss of generality, I assume that each consumer has an equity interest in one firm. The consumer's claims on the firm are a mixture of a one-period debt claim and a residual equity claim. The consumer makes all decisions for the firm, including the level of capital and the amount of debt. In the absence of government guarantees on the debt, the firm satisfies the Modigliani–Miller property of indifference to the mix of debt and equity.

7.3.3 Government

The government guarantees debt secured by capital. If the return next period is sufficiently negative to make capital less than the debt, the government pays the shortfall. The payout is the difference between the amount of debt and the amount of the collateral capital—the usual amount of a government payment for defaulted guaranteed debt or the cost of a bailout. The guarantee includes interest due on the debt, provided the interest rate is the economy's rate for default-free one-year obligations.

The government enforces capital requirements to limit its exposure. At the beginning of a period, the consumer invests debt and equity in the firm, which the firm uses to buy capital. The standard capital requirement limits debt to a fraction $1 - \alpha$ of the value of the capital held. But there is an exception and an exception to the exception. The exception is that firms may have higher leverage if the debt they bring into a period exceeds the fraction $1 - \alpha$ of the value of the capital. In that case they may keep debt at its previous level but may not take on any additional debt. The government forbears action against still-solvent firms that are in violation of the capital requirement. The exception to the exception is that debt may never exceed the value of the capital—the firm must be solvent during the period.

In principle, capital requirements for firms selling guaranteed debt often have the objective of disciplining firms that are solvent but lack all of the required capital, according to the doctrine of prompt corrective action (see Kocherlakota and Shim (2007) for an analysis of the doctrine that covers rather different issues from this chapter). The obstacles to enforcing marking to market are serious, however. Asset valuations tend to use historical rather than market values. If the government pushes guaranteed borrowers to mark their portfolios to market, the borrowers shift to assets that defy reliable valuation. Financial institutions are remarkably willing and able to create these assets, as recent experience has shown.

When a firm becomes insolvent because a decline in the asset value lowers the value of capital below the value of debt, the government makes its payoff. The firm starts the next year without any legacy debt.

Its new debt is constrained by the normal capital requirement.

The Modigliani–Miller property applies to any non-guaranteed debt the firm might issue. Such debt would have no effect on allocations in the model. For simplicity, I assume that firms issue only guaranteed debt.

In addition to capital requirements, I assume that the government can prevent specialization among firms that increases the government's exposure to payouts on debt guarantees. The danger is that some firms will pay out dividends to the point of borderline current solvency and that consumers, having reached the point where they prefer not to consume the dividends, invest them in another group of firms that are debt-free. I constrain all firms to have the same capital structures. The model would behave much the same way if specialization were permitted, but I exclude it to simplify the exposition.

The government finances the payments to honor its debt guarantees from a lump-sum tax, τ. The effects from the guarantee are accordingly pure substitution effects. The government balances its budget separately for each value of the state variables of the model and for each realization of the random return to capital.

7.3.4 Flows and Returns

Debt pays an interest rate of r_d. I discuss the determination of this rate in section 7.5.1. Because the government guarantee of debt is a giveaway, the consumer always lends the maximum permissible amount, which I write as

$$D = \min(K, \max(\hat{D}, (1 - \alpha)K)). \qquad (7.5)$$

Here D is the amount to be repaid, including interest and $\hat{D}$ is the legacy from last year's debt that might permit a higher level of debt to be held this year. The quantity $K = \hat{K} - c$ is the value of the capital the firm will carry through the period, The outside min enforces the solvency requirement during the year, the exception to the exception. The quantity $(1-\alpha)K$ is the standard capital requirement, but the max grants the exception for legacy shadow debt, $\hat{D}$. The consumer invests $D/(1+r_d)$ at the beginning of the year and receives D at the end of the year. Note that the capital requirement applies to the interest to be paid as well as the principal.

The consumer holds equity, Q, in the amount

$$Q = K - \frac{D}{1+r_d} \qquad (7.6)$$

at the beginning of the year, to make up the total assets of the firm, K, at the beginning of the year. The return the consumer earns on the equity investment is

$$\max(AK - D, 0). \qquad (7.7)$$

This is the residual after payment of interest and principal.

7.3.5 Consolidating Consumers and Firms

To study the allocations in the model, I consolidate each consumer with the corresponding firm. The decisions the consumer makes directly as the manager of the consolidated entity are the same as those that would occur under any efficient alternative managerial arrangement that operated the firm on behalf of its stakeholders. The allocations in the economy are the same when each firm is affiliated with one consumer as they would be if firms and consumers participated in capital and product markets, because of constant returns.

If the government did not guarantee debt, consumers would solve the dynamic program of equation (7.4). The debt guarantee gives the consumer an opportunity to capture additional value from the government payoff for default. The consolidated entity does not care about default itself, as it is a gain to the firm just offset by the loss to the consumer, but it does gain from the payment that the government makes when guaranteed debt defaults. Suppose first that the return ratio A is enough to pay off all the debt and interest. Then the combined return to the consumer is AK, the sum of the face value of debt and interest, D, and the return to equity in equation (7.7). Now suppose that the return is low enough that default occurs so that equity gets nothing and debt receives D in the form of the value of the capital AK and a guarantee payment $G = D - AK$. The consumer's combined return is $AK + G$. The only substantive effect of the guarantee is to add value not otherwise attainable when the asset value falls enough to trigger default. The optimizing consumer arranges to capture this value by taking on as much debt as possible. The availability of the subsidy has striking effects on the consumer's incentives to save and consume.

7.3.6 Consumer Decision Making

A consumer has two state variables, total debt $\hat{D}$ and the quantity of capital $\hat{K}$. At the beginning of a period, the consumer chooses a level of debt, D, to apply during the period and a level of consumption, c. The consumer always lends the maximum permissible amount in equation (7.5). The consumer invests $K = \hat{K} - c$ in the firm for the period.

The indicator variable z' takes the value 1 if default occurs next period and 0 if not. The law of motion for capital is

$$\hat{K}' = (1 - z')AK + z'D - \tau\left(\frac{\hat{D}}{\hat{K}}, A\right)\hat{K}. \tag{7.8}$$

Under solvency in the next period, the consumer earns the return AK for the capital K held during the period. Under insolvency, the new quantity of capital is the defaulted debt—the government makes up the difference between the actual amount of capital and the real value of the debt, so total capital is the real value of the debt. The tax payment $\tau(\hat{D}/\hat{K}, A)\hat{K}$ is a function of the state variables and the realization of the return, A, so that the government's budget can be balanced in each possible outcome.

The law of motion for the debt legacy is

$$\hat{D}' = (1 - z')D. \tag{7.9}$$

Legacy debt at the beginning of a period is the amount held through the earlier period unless the consumer defaults, in which case it becomes 0.

As in the simple growth model, capital enters the value function as $\hat{K}^{1-\gamma}$. The consumer's dynamic program is

$$V\left(\frac{\hat{D}}{\hat{K}}\right)\hat{K}^{1-\gamma} = \max_c \left(\frac{c^{1-\gamma}}{1-\gamma} + \mathbb{E}\,\beta V\left(\frac{\hat{D}'}{\hat{K}'}\right)\hat{K}'^{1-\gamma}\right). \tag{7.10}$$

7.4 Calibration

I use the value $\gamma = 2$ for the coefficient of relative risk aversion, as in chapter 2, and $\beta = 0.95$ for the

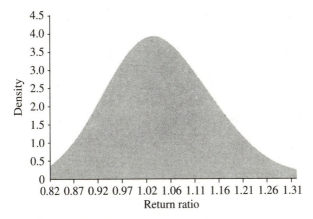

Figure 7.2. Density of annual return ratio.

discount ratio. I take the annual return ratio of capital to be log normal with log-mean $\alpha = 0.04$ and log-standard deviation $\sigma = 0.1$, truncated at the 1st and 99th percentiles. Figure 7.2 shows the density of the distribution. I take the capital requirement to be $\alpha = 30\%$.

7.5 Equilibrium

I will generally discuss the equilibrium of the model in terms of the ratio of consumption to the capital stock, $c/\hat{K}$, and the ratio of debt to the value of the capital stock, $\hat{D}/\hat{K}$. Recall that in the basic growth model without debt guarantees, $c/\hat{K}$ is constant.

At a time when legacy debt is low—just after default or following a positive return—the consumer will choose a value of debt controlled by the $(1 - \alpha)K$ in equation (7.5). If the return is positive in the next period, the consumer faces the same constraint and makes the same choice. If the return is negative, the consumer becomes subject to the exception of keeping the old value of debt, $\hat{D}$. The ratio $\hat{D}/\hat{K}$, the consumer's debt-related state variable, is correspondingly higher. Because the consumer is closer to insolvency, the probability of default next period rises. This change triggers the key behavioral response of the model. The price of consumption is lower when the probability of default is higher. Writing out equation (7.10) as

$$V\left(\frac{\hat{D}}{\hat{K}}\right)\hat{K}^{1-\gamma} = \max_c \frac{c^{1-\gamma}}{1-\gamma}$$

$$+ \mathbb{E}_A \beta\left[(1-z')V\left(\frac{\hat{D}'}{\hat{K}'}\right)(\hat{K}-c-\tau)^{1-\gamma}\right.$$

$$\left. + z'V(0)D'^{1-\gamma}\right]$$

$$(7.11)$$

shows that when $z' = 1$ and the consumer defaults, consumption turns out to have been free on the margin. The understanding of the possibility that consumption will turn out to be free results in a higher consumption choice before z' is realized. As random negative returns push the consumer into higher values of $\hat{D}/\hat{K}$, consumption rises.

The consumer's Euler equation is instructive on this point. Suppose that neither the regular capital requirement nor the exception to the exception capital requirement is binding, so the exception permitting the carrying forward of legacy debt is in effect—see equation (7.5). Consider the standard variational argument where the consumer reduces current consumption by a small amount, saves that amount for one year, and then consumes that amount plus its earnings for the year. This variation has no implications for the level of debt. Let c' be the level of consumption next year when the return A is known. Also let $f(A)$ be the density of the return ratio for capital. Then the Euler equation is

$$\beta \int_{A^*}^{\infty} c'^{-\gamma} A f(A) \, dA = c^{-\gamma}. \tag{7.12}$$

Here A^* is the value of the return ratio separating solvency from default. The Euler equation differs from the standard one only in the truncation of the integral on the left side. The omission of part of the distribution of marginal utility on the left, for the cases where the bailout makes consumption free, results in a lower current marginal utility and thus a higher level of current consumption. The consumer chooses high current consumption and plans that consumption will fall in the next year, on the average.

To find an equilibrium of the model, I approximate the consumer's value function $V(\hat{D}/\hat{K})$ as piecewise linear on a lattice of 200 knots of values of $\hat{D}/\hat{K}$. I approximate the truncated log-normal distribution of A by assigning probability 0.01 to each of the quantiles of the distribution running from $1/101$ to $100/101$, the truncation points. I start with an arbitrary tax function defined as a matrix of values whose rows correspond to the lattice values of the state variable and whose columns correspond to the 100 values of A. Then I solve the dynamic program given the tax function, by value-function iteration. Next I update the tax function to be the amounts of the guarantee paid out in each state and realization of A. I iterate this process to convergence, which is rapid.

The model implies a Markov transition process for the financial position of the consumer as measured by $\hat{D}/\hat{K}$. Figure 7.3 shows the stationary distribution

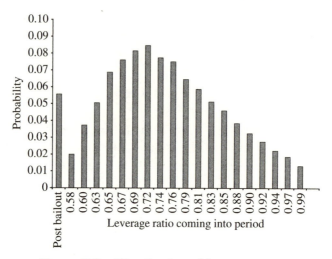

Figure 7.3.   Distribution of leverage ratio.

of the process. I approximate the process by a 200-
state version and then aggregate to the bins shown in
the figure. About half the time, $\hat{D}/\hat{K}$ is at or below the
notional limit of 70%. Although consumers will move
up to the 70% level in any period when $\hat{D}/\hat{K}$ falls below 70%, positive returns are common and large, so
the state variable is often below 70% because the most
recent return was positive. For 5.5% of the time, consumers have no legacy debt because they defaulted
in the previous period. For the other half of the time,
consumers have higher leverage than 70% because the
cumulative return to capital is below its previous peak
and the legacy exception permits the extra debt.

Figure 7.4 shows the consumer's choice of debt for
the coming period, D, which I normalize as $D/\hat{K}$. Much
of the time the choice is $(1 - \alpha)K$, the 70% mentioned
above. This is the choice of a consumer just past insolvency and of a consumer who is not eligible for

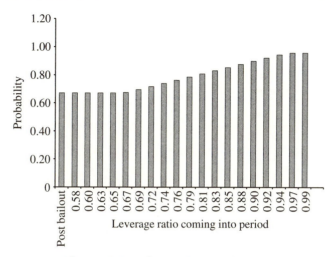

Figure 7.4. Chosen leverage ratio.

the legacy exception. Whenever the return is negative, the consumer is eligible for the legacy exception, either as a continuation and extension of the exception or for the first time. In years following multiple negative returns, consumers may enjoy values of the leverage ratio $D/\hat{K}$ close to its upper limit of almost exactly 1—the exception to the exception, where $D/\hat{K} = (1 - c/\hat{K})$.

Figure 7.5 shows the probability of default conditional on the leverage ratio. It rises monotonically to its highest possible level of 35%.

Figure 7.6 shows how consumption responds to the state of the economy. In the region of the legacy exception ($0.703 \leqslant \hat{D}/\hat{K} \leqslant 0.987$) consumption rises with the leverage ratio because the probability of default rises, making consumption this period cheaper because the government finances it when default occurs. At the highest values of the debt ratio, consumption is

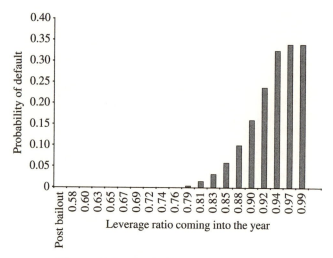

Figure 7.5. Probability of default as
a function of the leverage ratio.

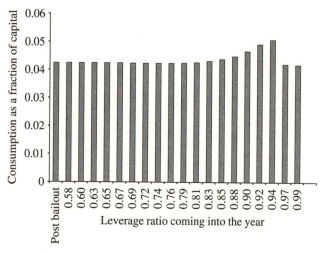

Figure 7.6. Consumption/capital ratio as
a function of the leverage ratio.

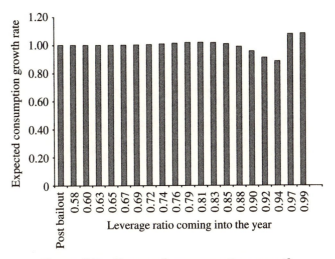

Figure 7.7. Expected consumption growth
rate as a function of the leverage ratio.

much lower, because of the exception to the exception
requiring solvency during the period.

7.5.1 Interest Rate on Debt

The interest rate on debt, r_d, satisfies the consumption
CAPM asset-pricing condition,

$$(1 + r_d) \beta \, \mathbb{E}_A \left(\frac{c'}{c} \right)^{-y} = 1. \tag{7.13}$$

As usual in the consumption CAPM, the rate varies
positively with expected consumption growth. It also
varies because consumption growth covaries with A
(the conditional distribution of A is invariant but
the conditional distribution of c'/c varies by current state). Figure 7.7 shows expected consumption
growth as a function of the financial condition of the
consumer. Consumption hardly changes at all when

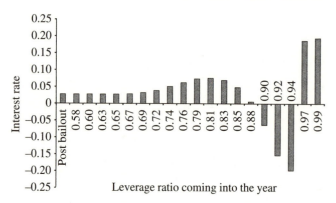

Figure 7.8. Interest rate as
a function of the leverage ratio.

the consumer is in the post-default state, so the expected growth rate in that state is essentially zero. Consumers near or below their capital requirements expect consumption to grow, as there is a chance that they will move next year into a more leveraged position where a higher level of consumption is optimal. But once consumers reach higher leverage, they face the high likelihood of default with a sharp drop of consumption back to the normal leverage level, so expected consumption growth is negative.

Figure 7.8 shows the interest rate on guaranteed debt as a function of leverage. When leverage is low, the rate is 13%: 5% real return to capital plus 4% expected inflation plus about 4% from expected consumption growth. As long as leverage does not exceed 85%, the rate remains above 3%. For higher amounts of leverage, the force of expected declines in consumption takes over. Above 90% leverage, the rate becomes negative as participants know that default and the accompanying resumption of the normal, lower level of consumption is fairly likely (see figure 7.5).

7.5.2 Return to Equity

The equity investment $Q = pK - D/(1 + r_d)$ in one period pays off

$$(1 + r_Q)Q = \max(AK - D, 0). \qquad (7.14)$$

The return satisfies the pricing condition,

$$\mathbb{E}_A\left[(1 + r_Q)\beta\left(\frac{c'}{c}\right)^{-\gamma}\right] = 1. \qquad (7.15)$$

7.5.3 Financial Flows

In one year, after consuming the amount c, the owner lends the firm $D/(1 + r_d)$ and provides equity Q which adds up to capital worth K in the hands of the firm. Production AK occurs before the next year. The firm pays this amount to its owner as the return of borrowed funds, interest on those funds at rate r_d, and the payout of the owner's residual equity, if the firm has not defaulted. The government makes up the difference to the owner if AK falls short of D. The model generates matrices of flows, indexed by the state of the economy in the first year, $\hat{D}/\hat{K}$, and by the random return A. I describe these matrices in two ways.

First, figure 7.9 shows the expectation of the ratio of the flows to the original value of capital, $\hat{K}$. The horizontal axis shows $\hat{D}/\hat{K}$, the debt legacy, as a ratio to the value of capital brought into the period. If the legacy leverage ratio is less than the notional upper limit of $1 - \alpha$, 70%, the consumer lends the firm 70% of the value of the capital, $(1 - \alpha)(\hat{K} - c)$. Default has a low probability, so the firm is able to repay the debt with interest, as shown in the top line, and to return the consumer's equity at a level with a reasonable return. The expected government guarantee

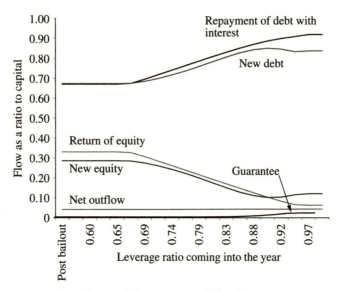

Figure 7.9. Expected flows as
functions of the leverage ratio.

payment, shown at the bottom, is 0. The firm issues
new debt and new equity with values somewhat below the amounts returned, reflecting the interest and
return to equity from the earlier investments. The expected net outflow, shown close to the bottom, is the
expected value of the consumer's consumption at the
beginning of the new period.

On the right side of figure 7.9, the consumer benefits from the exception for legacy debt. Earlier negative
returns have left the consumer with the right to issue more debt to replace the high level of legacy debt.
The rising lines show the expected repayment of debt
including interest and the expected issuance of new
debt. Their upward slope reflects the higher value of
legacy debt. The bottom line shows the government's
expected guarantee payoff. Expected return of equity

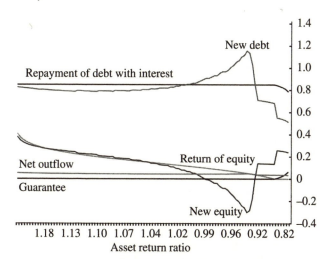

Figure 7.10. Flows as functions of the asset return ratio when prior leverage is 0.85.

and issuance of new equity decline with leverage. Equity depletion is visible as the excess of expected new debt over repayment of debt and the shortfall of new equity over the return of equity. Notice that the net outflow, equal to consumption, is essentially flat. Here, consumption next year is stated as a ratio to the value brought into this year; it is not next year's consumption/capital ratio, which is shown in figure 7.6. The rise in the consumption/capital ratio that occurs when the consumer moves closer to default is largely offset by the fact that the move occurs when returns are negative, which lowers next year's capital. The ratio rises, but the actual amount of consumption remains about the same.

The second view of the matrix of flows is along one row, showing the various possible outcomes at a given level of prior leverage. Figure 7.10 plots the flows with the asset return ratio on the horizontal axis and the flows as fractions of the earlier value of capital on the vertical axis, for $\hat{D}/\hat{K} = 0.85$. For consistency with the other plots in the chapter, where declines in asset values move the economy to the right, the horizontal axis has a high new value on the left and a low new value on the right. The left side of the figure shows what happens when the economy retreats from high leverage because the asset value rises. No default occurs, so all debt is repaid with interest. Newly issued debt is about equal to debt repayments. For modest negative returns, the interest rate turns negative because of the impending decline in consumption—see figure 7.8. Now an obligation to repay principal and interest at the end of the year yields more than a dollar of current funds for each dollar returned at the end of the year. New debt issuance rises to a peak, offset by a reduction in equity. Firms are borrowing heavily and paying out the proceeds as dividends or share repurchases. Consumption remains constant, but the consumption/capital ratio rises, as shown in figure 7.6. Farther to the right of figure 7.10, default occurs. Repayment of debt drops because the firm's collateral has fallen below the face value of debt and interest. The guarantee payment rises. New debt is below the level on the left, because the firm loses its legacy right to issue more debt upon default and is limited by its now lower amount of capital. Issuance of equity makes up the difference.

7.5.4 Equity Depletion

Equity depletion occurs when a decline in the asset value causes the probability of default to rise. Figure 7.10 shows how depletion occurs. If the asset value falls, the firm pays out a substantial amount of equity. Most is replaced by corresponding debt investment. Two factors account for the equity outflow that is replaced by debt. First is the compelling advantage of leverage when the government guarantees debt. From the joint perspective of the firm and investor, debt is unambiguously a good thing and the firm is always at a corner solution with maximum permissible guaranteed debt. A fall in the asset value relaxes the debt constraint because it lowers the interest rate. The firm immediately borrows up to the new, higher limit. Recall that the amount borrowed is $D/(1 + r_d)$. The firm then pays out the difference as dividends and thereby reduces equity. The second factor is that investors reduce their total funding when the asset value falls, because they increase consumption. The new level of funding to the firm is $(\hat{K}' - c')$ and c' rises when A' is less than 1. Although the decline in total funds available is small relative to the shift from equity to debt, the increase in consumption is critical to the process, because the rise in consumption as firms approach default creates a decrease in expected consumption growth and thus the lower interest rate that permits the expansion of debt.

7.6 Roles of Key Parameters

Figure 7.11 compares economies with different values of the key parameters, in terms of the way consumption depends on the leverage state variable, $\hat{D}/(p\hat{K})$. The first case is a capital requirement of $\alpha = 0.1$ instead of 0.3 as in the base case. Firms coming into the year with leverage less than 90% immediately issue debt up to that point. The behavior of consumption is

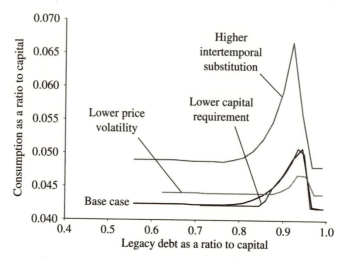

Figure 7.11. Comparison of consumption as
a function of leverage for four cases.

different from the base case only in the region where
the base-case consumer is levered beyond the basic
70% level but less than the 90% level that is advan-
tageous to the consumer in the economy with lower
capital requirements. In interpreting this finding, one
should keep in mind that these are substitution effects. The actual amounts of guarantee payoffs are
much higher with the lower capital requirement, but
these are taxed away.

The second case is less risk aversion and higher intertemporal substitution—specifically, a value of the
curvature parameter y of 1.1 in place of its value of 2
in the base case. Higher substitution greatly increases
the volatility of consumption. Consumers concentrate
their spending in times of cheap consumption to a
greater extent, so consumption is lower when leverage is at or below the capital requirement and higher

when leverage is higher and the probability of default is higher, so consumption is cheaper in expectation. Notice that the consumption/capital ratio is always higher for the economy with more intertemporal substitution. As a result, its endogenous growth rate is lower.

The third case is lower volatility of the return ratio. I reduce the standard deviation of the return ratio, A, to 0.05. The result is, as expected, lower volatility of the consumption/capital ratio.

Volatility in other variables follows closely from the volatility in consumption shown in figure 7.11. In particular, movements in the interest rate track expected consumption growth, which is more volatile when intertemporal substitution is higher (y is lower), but less volatile with lower asset-return volatility.

7.7 Concluding Remarks

The essential feature of a government guarantee of debt that yields the results in this chapter is the government's failure to take prompt corrective action. Asset-value declines permit higher leverage, with a higher probability of government payoff. The consumption bulge that is the source of the volatility would not occur if the government insisted on equity contributions to make up for asset-value declines so as to keep leverage constant. Thus the relevance of the basic mechanism studied in this chapter to modern economic instability rests on the government's failure to measure the market value of the collateral backing guaranteed debt and the government's resulting failure to require equity infusions following negative returns. My impression is that these failures are the rule. Many organizations enjoying effective guarantees, such as investment banks, do not even have stated capital requirements. Financial intermediaries have shown great ingenuity in creating instruments that defy current valuation.

Government guarantees are only one of many issues raised by recent financial events. The point of this chapter is to pursue the effects of guarantees in a dynamic general-equilibrium setting. Guarantees introduce volatility in key variables in an economy that would otherwise evolve smoothly. In particular, if the government administers guarantees in terms of nominal measures, then otherwise neutral nominal developments have profound real effects.

7.8 Appendix: Value Functions

Proposition 7.1. *The value function for the basic growth model has the form*

$$\bar{V} K^{1-\gamma}. \tag{7.16}$$

Proof. If a function satisfies

$$V(K) = \max_{c} \frac{c^{1-\gamma}}{1-\gamma} + V((1+r)(K-c)), \tag{7.17}$$

it is the unique value function (see Mas-Colell et al. 1995, p. 969). Suppose that $\bar{V}$ satisfies

$$\bar{V} = \max_{\tilde{c}} \frac{\tilde{c}^{1-\gamma}}{1-\gamma} + \beta \bar{V}[(1+r)(1-\tilde{c})]^{1-\gamma}. \tag{7.18}$$

Then

$$\bar{V} K^{1-\gamma} = \max_{\tilde{c}} \frac{(\tilde{c}K)^{1-\gamma}}{1-\gamma} + \beta \bar{V}[(1+r)(K-\tilde{c}K)]^{1-\gamma}. \tag{7.19}$$

Because maximization over $\tilde{c}$ is equivalent to maximizing over c, equation (7.19) is equivalent to (7.17). □

Proposition 7.2. *The value function for the model with guaranteed shadow debt has the form*

$$\bar{V}_0\left(\frac{D}{pK}\right)K^{1-y} \quad and \quad \bar{V}_1 K^{1-y}. \qquad (7.20)$$

Proof. Suppose that $\bar{V}_0(D/pK)$ satisfies

$$\bar{V}_0\left(\frac{D}{pK}\right) = \max_{\tilde{c}} \frac{\tilde{c}^{1-y}}{1-y} + \mathbb{E}\frac{F_t}{1+r}, \qquad (7.21)$$

where

$$F_t = (1-z')\bar{V}_0\left(\frac{D'}{p'K'}\right)[(1+r)(1-\tilde{c})]^{1-y}$$

$$+ z'\bar{V}_1\left(\frac{\hat{D}}{pK}\frac{p}{p'} - T'\right)^{1-y}. \qquad (7.22)$$

Note that if the consumer remains solvent, $z' = 0$,

$$(1+r)(1-\tilde{c}) = \frac{K'}{K} \qquad (7.23)$$

while if the consumer defaults,

$$\frac{\hat{D} - T'}{pK}\frac{p}{p'} = \frac{K'}{K}. \qquad (7.24)$$

Substituting these two equations into (7.11) and multiplying by K^{1-y}, I find

$$\bar{V}_0\left(\frac{D}{pK}\right) = \max_{\tilde{c}} \frac{\tilde{c}^{1-y}}{1-y}$$

$$+ \mathbb{E}\frac{1}{1+r}\left[(1-z')\bar{V}_0\left(\frac{D'}{p'K'}\right)K'^{1-y}\right.$$

$$\left. + z'\bar{V}_1 K'^{1-y}\right], \qquad (7.25)$$

which is the same as the Bellman equation (7.10) when $z = 0$. The demonstration for $z = 1$ follows the same principles as in proposition 7.1. □

References

Aguiar, Mark, and Erik Hurst. 2005. Consumption versus expenditure. *Journal of Political Economy* **113**:919-48.

Akerlof, George A., and Paul M. Romer. 1993. Looting: the economic underworld of bankruptcy for profit. Brookings Papers on Economic Activity, pp. 1-73.

Attanasio, Orazio P., and Hamish Low. 2004. Estimating Euler equations. *Review of Economic Dynamics* **7**:406-35.

Attanasio, Orazio P., and Guglielmo Weber. 1993. Consumption growth, the interest rate, and aggregation. *Review of Economic Studies* **60**:631-49.

———. 1995. Is consumption growth consistent with intertemporal optimization? Evidence from the consumer expenditure survey. *Journal of Political Economy* **103**:1121-57.

Attanasio, Orazio P., James Banks, Costas Meghir, and Guglielmo Weber. 1999. Humps and bumps in lifetime consumption. *Journal of Business & Economic Statistics* **17**:22-35.

Banks, James, Richard Blundell, and Sarah Tanner. 1998. Is there a retirement-savings puzzle? *American Economic Review* **88**: 769-88.

Barsky, Robert B., F. Thomas Juster, Miles S. Kimball, and Matthew D. Shapiro. 1997. Preference parameters and behavioral heterogeneity: an experimental approach in the health and retirement study. *Quarterly Journal of Economics* **112**:537-79.

Black, Fischer, and Myron Scholes. 1973. The pricing of options and corporate liabilities. *Journal of Political Economy* **81**:637-54.

Blundell, Richard, Luigi Pistaferri, and Ian Preston. 2008. Consumption inequality and partial insurance. *American Economic Review* **98**:1887-921.

Brown, Jeffrey R., and Amy N. Finkelstein. 2008. The interaction of public and private insurance: Medicaid and the long-term care insurance market. *American Econmic Review* **98**: 1083-102.

Browning, Martin, and Thomas F. Crossley. 2001. Unemployment insurance benefit levels and consumption changes. *Journal of Public Economics* **80**:1-23.

Browning, Martin, Angus Deaton, and Margaret Irish. 1985. A profitable approach to labor supply and commodity demands over the life-cycle. *Econometrica* **53**:503–44.

Campbell, John Y., and John H. Cochrane. 1999. By force of habit: a consumption-based explanation of aggregate stock market behavior. *Journal of Political Economy* **107**:205–51.

Carroll, Christopher D. 2001. Death to the log-linearized consumption Euler equation! (And very poor health to the second-order approximation.) *Advances in Macroeconomics* **1**(1), Article 6.

Chetty, Raj. 2006. A new method of estimating risk aversion. *American Economic Review* **96**:1821–34.

Cohen, Alma, and Liran Einav. 2007. Estimating risk preferences from deductible choice. *American Economic Review* **97**:745–88.

Costa, Dora, and Matthew Kahn. 2004. Changes in the value of life, 1940–1980. *Journal of Risk and Uncertainty* **29**:159–80.

Domeij, David, and Martin Floden. 2006. The labor-supply elasticity and borrowing constraints: why estimates are biased. *Review of Economic Dynamics* **9**:242–62.

Dow, William H., Tomas J. Philipson, and Xavier Sala-i-Martin. 1999. Longevity complementarities under competing risks. *American Economic Review* **89**:1358–71.

Fisher, Jonathan, David S. Johnson, Joseph Marchand, Timothy M. Smeeding, and Barbara Boyle Terrey. 2005. The retirement consumption conundrum: evidence from a consumption survey. Center for Retirement Research at Boston College, Working Paper 1005-14.

Gerdtham, Ulf-G., and Bengt Jonsson. 2000. International comparisons of health expenditure: theory, data and econometric analysis. In *Handbook of Health Economics* (ed. Anthony J. Culyer and Joseph P. Newhouse). Amsterdam: North-Holland.

Guvenen, Fatih. 2006. Reconciling conflicting evidence on the elasticity of intertemporal substitution: a macroeconomic perspective. *Journal of Monetary Economics* **53**:1451–72.

Hall, Robert E. 2008. Cyclical movements along the labor supply function. In *Labor Supply in the New Century* (ed. Christopher L. Foote, Katherine Bradbury, and Robert K. Triest), pp. 241–64. Federal Reserve Bank of Boston.

——. 2009. Reconciling cyclical movements in the marginal value of time and the marginal product of labor. *Journal of Political Economy* **117**:281–323.

Hall, Robert E., and Charles I. Jones. 2007. The value of life and the rise in health spending. *Quarterly Journal of Economics* **122**:39–72.

Hall, Robert E., and Paul R. Milgrom. 2008. The limited influence of unemployment on the wage bargain. *American Economic Review* **98**:1653–74.

Hall, Robert E., and Susan E. Woodward. 2007. The incentives to start new companies: evidence from venture capital. Hoover Institution, Stanford University. (Available at stanford.edu/~rehall.)

———. Forthcoming. The burden of the idiosyncratic risk of entrepreneurship. *American Econmic Review*.

Hammitt, James K., Jin-Tan Liu, and Jin-Long Liu. 2000. Survival is a luxury good: the increasing value of a statistical life. Mimeo, Harvard University.

Hansen, Lars Peter, and Ravi Jagannathan. 1991. Implications of security market data for models of dynamic economies. *Journal of Political Economy* **99**:225–62.

Hurst, Erik. 2008. The retirement of a consumption puzzle. NBER Working Paper 13789.

Imbens, Guido W., Donald B. Rubin, and Bruce I. Sacerdote. 2001. Estimating the effect of unearned income on labor earnings, savings, and consumption: evidence from a survey of lottery players. *American Econmic Review* **91**:778–94.

Jones, Charles I. 2004. Why have health expenditures as a share of GDP risen so much? Mimeo, University of California, Berkeley.

Judd, Kenneth. 1998. *Numerical Methods in Economics*. Cambridge, MA: MIT Press.

Kimball, Miles S., and Matthew D. Shapiro. 2003. Labor supply: are the income and substitution effects both large or both small? Department of Economics, University of Michigan.

Kocherlakota, Narayana, and Ilhyock Shim. 2007. Forbearance and prompt corrective action. *Journal of Money, Credit, and Banking* **39**:1107–29.

Low, Hamish, Costas Meghir, and Luigi Pistaferri. 2008. Wage risk and employment risk over the life cycle. Institute for Fiscal Studies, London.

Mas-Colell, Andreu, Michael D. Whinston, and Jerry R. Green. 1995. *Microeconomic Theory*. Oxford University Press.

Mehra, Rajnish, and Edward C. Prescott. 1985. The equity premium: a puzzle. *Journal of Monetary Economics* **15**:145–61.

Merz, Monika. 1995. Search in the labor market and the real business cycle. *Journal of Monetary Economics* **36**:269-300.

Miniaci, Raffaele, Chiara Monfardini, and Guglielmo Weber. 2003. Is there a retirement consumption puzzle in Italy? Institute for Fiscal Studies, Working Paper 03/14.

Mortensen, Dale T., and Christopher Pissarides. 1994. Job creation and job destruction in the theory of unemployment. *Review of Economic Studies* **61**:397-415.

Mulligan, Casey B. 1998. Substitution over time: another look at life-cycle labor supply. *NBER Macroeconomics Annual*, pp. 75-134.

Newhouse, Joseph P. 1992. Medical care costs: how much welfare loss? *Journal of Economic Perspectives* **6**:3-21.

Pistaferri, Luigi. 2003. Anticipated and unanticipated wage changes, wage risk, and intertemporal labor supply. *Journal of Labor Economics* **21**:729-54.

Rogerson, Richard, and Johanna Wallenius. Forthcoming. Micro and macro elasticities in a life cycle model with taxes. *Journal of Economic Theory*.

Rosen, Sherwin. 1988. The value of changes in life expectancy. *Journal of Risk and Uncertainty* **1**:285-304.

Shimer, Robert. 2005. The cyclical behavior of equilibrium unemployment and vacancies. *American Economic Review* **95**:24-49.

———. 2008. Convergence in macroeconomics: the labor wedge. *American Economic Journal: Macroeconomics* **1**:280-97.

Stokey, Nancy L., and Robert E. Lucas. 1989. *Recursive Methods in Economic Dynamics* (with Edward C. Prescott). Cambridge, MA: Harvard University Press.

Weisbrod, Burton A. 1991. The health care quadrilemma: an essay on technological change, insurance, quality of care, and cost containment. *Journal of Economic Literature* **29**:523-52.

Index

.

The Gorman Lectures
in Economics

Richard Blundell, Series Editor

Terence (W. M.) Gorman was one of the most distinguished economists of the twentieth century. His ideas are so ingrained in modern economics that we use them daily with almost no acknowledgment. The relationship between individual behavior and aggregate outcomes, two-stage budgeting in individual decision making, the "characteristics" model which lies at the heart of modern consumer economics, and a conceptual framework for "adult equivalence scales" are but a few of these. For over fifty years he guided students and colleagues alike in how best to model economic activities as well as how to test these models once formulated.

During the late 1980s and early 1990s Gorman was a Visiting Professor of Economics at University College London. He became a key part of the newly formed and lively research group at UCL and at the Institute for Fiscal Studies. The aim of this research was to avoid the obsessive labeling that had pigeonholed much of economics and to introduce a free flow of ideas between economic theory, econometrics, and empirical evidence. It worked marvelously and formed the mainstay of economics research in the Economics Department at UCL. These lectures are a tribute to his legacy.

Terence had a lasting impact on all who interacted with him during that period. He was not only an active and innovative economist but he was also a dedicated teacher and mentor to students and junior colleagues. He was generous with his time and more than one discussion with Terence appeared later as a scholarly article inspired by that conversation. He used his skill in mathematics as a framework for his approach but he never insisted on that. What was essential was coherent and logical understanding of economics.

Gorman passed away in January 2003, shortly after the second of these lectures. He will be missed but his written works remain to remind all of us that we are sitting on the shoulders of a giant.

Richard Blundell, University College London and
Institute for Fiscal Studies

Biography

Gorman graduated from Trinity College, Dublin, in 1948 in Economics and in 1949 in Mathematics. From 1949 to 1962 he taught in the Commerce Faculty at the University of Birmingham. He held Chairs in Economics at Oxford from 1962 to 1967 and at the London School of Economics from 1967 to 1979, after which he returned to Nuffield College Oxford as an Official Fellow. He remained there until his retirement. He was Visiting Professor of Economics at UCL from 1986 to 1996. Honorary Doctorates have been conferred upon him by the University of Southampton, the University of Birmingham, the National University of Ireland, and University College London. He was a Fellow of the British Academy, an honorary Fellow of Trinity College Dublin and of the London School of Economics, an honorary foreign member of the American Academy of Arts and Sciences and of the American Economic Association. He was a Fellow of the Econometric Society and served as its President in 1972.